PRODUCTION AND OPERATIONS MANAGEMENT

PRATIBHA GARG
B.E., M.B.A.
Assistant Professor,
S.S.I.E.T., Dera Bassi

S.K. KATARIA & SONS®
Publisher of Engineering Books
4885/109, Prakash Mahal, Dr. Subhash Bhargav Lane,
Opposite Delhi Medical Association, Daryaganj, New Delhi-110002
Phone: +91-11-43551243; Mobile: +91-9871775858, 7678425209
e-mail: katariabook@gmail.com, katariabooks@yahoo.com
Website: www.skkatariaandsons.com
(e-books, subscriptions and bulk buying also available on our website)

Published By :

S.K. KATARIA & SONS®

4885/109, Prakash Mahal, Dr. Subhash Bhargav Lane,
Opposite Delhi Medical Association, Daryaganj, New Delhi-110002
Phone: +91-11-43551243; Mobile: +91-9871775858, 7678425209
e-mail: katariabook@gmail.com, katariabooks@yahoo.com

STATUTORY WARNING

Information contained in this book has been obtained by author from sources believed to be reliable and are correct to the best of his knowledge. Every effort has been made to avoid errors and omissions and ensure accuracy. Any error or omission noted may be brought to the notice of the publisher which shall be taken care of in forthcoming edition of this book. However, neither the publisher nor the author guarantee the accuracy or completeness of any information published herein, and neither the publisher nor author take any responsibility of liability for any inconvenience, expenses, losses or damage to anyone resulting from contents of this book.

The book is meant for educational and learning purposes and there is no attempt on the part of publisher/author to render engineering and other professional services. If such services are required, the assistance of an appropriate professional should be sought.

The author of the book has taken all possible care to ensure that the contents of the book do not violate any existing copyright or other intellectual property rights of any person in any manner whatsoever. In the event, the author has been unable to track any source and if any copyright has been inadvertently infringed, the facts may be brought to the notice of the publisher in writing for corrective action.

Send all correspondence to : **M/s S.K. Kataria & Sons,** New Delhi
Email: katariabook@gmail.com

First Edition: January, 2012; Reprint 2026

Price: ₹ 195/-

Laser Typesetting at: Goswami Associates

Printed at: Arora Printing Press

Preface

This book is brought so as to give overview of the subject Production and Operations Management. Entire matter in the book is a copy of my lecture notes prepared while I was teaching. "Production and Operations Management" subject to M.B.A. classes. I hope this book will meet the long-felt need of students and teachers alike. Nothing would give me more satisfaction than its acceptance by them.

I would like to acknowledge the persons who have been instrumental either in motivating me constantly in my venture or in helping me at various occasions in finalizing my manuscript. In the beginnings I sincerely express my indebtedness to the management and all my colleagues at Sri Sukhmani Institute of Engineering and Technology, Darabassi. Next I must express my heartful gratitude to my parents, my in-laws and my dear brother Jonny who gave me their moral support all along the work. At the end I would like to mention with gratitude the contribution that I received from my husband who has always stood by me like a pillar with infinite patience and morel courage and particularly that of my little daughter Ashiya who bore patiently with my whims and petulances at time for beyond her age.

Contents

CHAPTER 1

Production and Operations Management

1.1 INTRODUCTION

Production/operations management is the process, which combines and transforms various resources used in the production/operations subsystem of the organization into value added product/services in a controlled manner as per the policies of the organization. Therefore, it is that part of an organization, which is concerned with the transformation of a range of inputs into the required (products/services) having the requisite quality level.

The set of interrelated management activities, which are involved in manufacturing certain products, is called as production management. If the same concept is extended to services management, then the corresponding set of management activities is called as operations management.

1.2 VARIOUS DEFINITION OF PRODUCTION AND OPERATION MANAGEMENT

(*i*) "Production management is the process of effectively planning and regulating that part of an enterprise which are responsible for the actual transformation of material into finished products". **—E.F.L. Brech**

(*ii*) "Production management is concerned with those processes which convert the inputs into the outputs. The inputs are various resources like raw materials, men, machines, methods, etc. and the outputs are good and services." **—H.A. Harding**

(*iii*) "Production Management deals with decision-making related to production process so that the resulting goods or service is produced according to specification in the amounts and by the schedule demanded and at minimum cost." **—E.S. Buffa**

1.3 EVOLUTION OF PRODUCTION AND OPERATIONS MANAGEMENT

For over two centuries operations and production management has been recognised as an important factor in a country's economic growth.

The traditional view of manufacturing management began in eighteenth century when **Adam Smith** recognised the economic benefits of specialisation of labour. He recommended breaking of jobs down into subtasks and recognises workers to specialised tasks in which they would become highly skilled and efficient. In the early twentieth century, F.W. Taylor implemented Smith's theories and developed scientific management. From then till 1930, many techniques were developed prevailing the traditional view. Brief information about the contributions to manufacturing management is shown in the Table 1.1.

Table 1.1: Historical summary of operations management

Data	*Contribution*	*Contributor*
1776	Specialization of labour in manufacturing	Adam Smith
1799	Interchangeable parts, cost accounting	Eli Whitney and others
1832	Division of labour by skill; assignment of jobs by skill; basics of time study	Charles Babbage
1900	Scientific management time study and work study developed; dividing planning and doing of work	Frederick W. Taylor
1900	Motion study of jobs	Frank B. Gilbreth
1901	Scheduling techniques for employees, machines, jobs in manufacturing	Henry L. Gantt
1915	Economic lot sizes for inventory control	F.W. Harris
1927	Human relations; the Hawthorne studies	Elton Mayo
1931	Statistical inference applied to product quality: quality control charts	W.A. Shewart
1935	Statistical sampling applied to quality control: inspection sampling plans	H.F. Dodge & H.G. Roming
1940	Operations research application in World War II	P.M. Blacker and others.
1946	Digital computer	John Mauchlly and J.P. Eckert
1947	Linear programming	G.B. Dantzig, Williams and others
1950	Mathematical programming, on-linear and stochastic processes	A. Charnes, W.W. Cooper and others
1951	Commercial digital computer: large-scale computations available.	Sperry Univac
1960	Organizational behaviour: continued study of people at work	L. Cummings, L. Porter

1970	Integrating operations into overall strategy and policy, Computer applications to manufacturing, Scheduling and control, Material requirement planning (MRP)	W. Skinner, J. Orlicky and G. Wright
1980	Quality and productivity applications from Japan: robotics, CAD-CAM	W.E. Deming and J. Juran.

Production management becomes the acceptable term from 1930s to 1950s. As F.W. Taylor's works become more widely known, managers developed techniques that focussed on economic efficiency in manufacturing. Workers were studied in great detail to eliminate wasteful efforts and achieve greater efficiency. At the same time, psychologists, socialists and other social scientists to study people and human behaviour in the working environment. In addition, economists, mathematicians, and computer socialists contributed newer, more sophisticated analytical approaches.

With the 1970s emerges two distinct changes in our views. The most obvious of these reflected in the new name **operations management** was a shift in the service and manufacturing sectors of the economy. As service sector became more prominent, the change from 'production' to 'operations' emphasized the broadening of our field to service organizations. The second, more suitable change was the beginning of an emphasis on synthesis, rather than just analysis, in management practices.

1.4 PRODUCTION SYSTEM

The production system of an organization is that part, which produces products of an organization. It is that activity whereby resources, flowing within a defined system, are combined and transformed in a controlled manner to add value in accordance with the policies communicated by management. A production system is very closely related with the environment. It takes from the environment-personnel, engineering and financial inputs in the form of human, capital, material, machines and information respectively. The system has transformation process which includes product design, process planning, production control and maintenance. The system gives outputs in the form of products and services to the environment. The production system takes the feedback from the environment, in the form of complaints and suggestions.

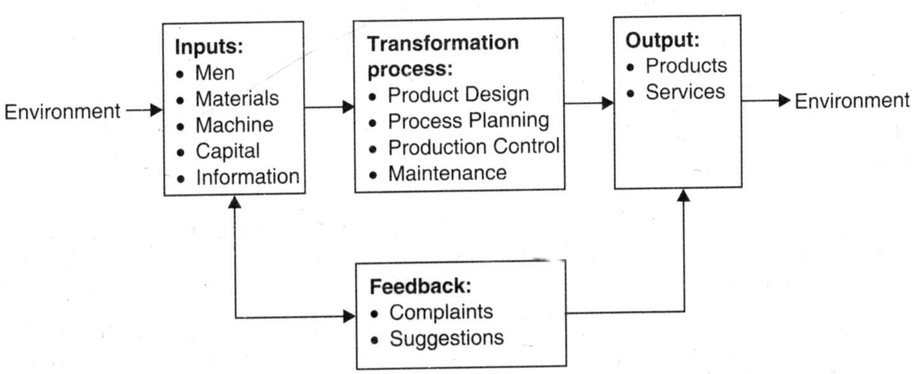

Fig. 1.1

1.4.1 Classification of Production System

Production system can be classified as job shop, Batch, Mass and continuous production system.

1. Job Shop Production

Job shop production is characterised by manufacturing of one or few quantity of products designed and produced as per the specification of customers within prefixed time and cost. The distinguishing feature of this is low volume and high variety of products. A job shop comprises of general purpose machines arranged into different departments. Each job demands unique technological requirements, demands processing on machines in a certain sequence.

Characteristics

(*i*) High variety of products and low volume.

(*ii*) Highly skilled operators who can take up each job as a challenge because of uniqueness.

(*iii*) Use of general purpose machines and facilities.

(*iv*) Large inventory of materials, tools, parts.

(*v*) Detailed planning is essential for sequencing the requirements of each product, capacities for each work centre and order priorities.

Advantages

Following are the advantages of job shop production:

1. Because of general purpose machines and facilities variety of products can be produced.
2. Operators will become more skilled and competent, as each job gives them learning opportunities.
3. Full potential of operators can be utilised.
4. Opportunity exists for creative methods and innovative ideas.

Limitations

Following are the limitations of job shop production:

1. Higher cost due to frequent set up changes.
2. Higher level of inventory at all levels and hence higher inventory cost.
3. Production planning is complicated.
4. Larger space requirements.

Batch Production

Batch production is defined by American Production and Inventory Control Society (APICS) "*as a form of manufacturing in which the job passes through the functional departments in lots or batches and each lot may have a different routing.*" It is characterised by the manufacture of limited number of products produced at regular intervals and stocked awaiting sales.

Characteristics

Batch production system is used under the following circumstances:

1. When there is shorter production runs.
2. When plant and machinery are flexible.

3. When plant and machinery set up is used for the production of item in a batch and change of set up is required for processing the next batch.
4. When manufacturing lead time and cost are lower as compared to job order production.

Advantages

Following are the advantages of batch production:

1. Better utilisation of plant and machinery.
2. Promotes functional specialisation.
3. Cost per unit is lower as compared to job order production.
4. Lower investment in plant and machinery.
5. Flexibility to accommodate and process number of products.
6. Job satisfaction exists for operators.

Limitations

Following are the limitations of batch production:

1. Material handling is complex because of irregular and longer flows.
2. Production planning and control is complex.
3. Work in process inventory is higher compared to continuous production.
4. Higher set up costs due to frequent changes in set up.

Mass Production

Manufacture of discrete parts or assemblies using a continuous process are called mass production. This production system is justified by very large volume of production. The machines are arranged in a line or product layout. Product and process standardisation exists and all outputs follow the same path.

Characteristics

Mass production is used under the following circumstances:

1. Standardisation of product and process sequence.
2. Dedicated special purpose machines having higher production capacities and output rates.
3. Large volume of products.
4. Shorter cycle time of production.
5. Lower in process inventory.
6. Perfectly balanced production lines.
7. Flow of materials, components and parts is continuous and without any back tracking.
8. Production planning and control is easy.
9. Material handling can be completely automatic.

Advantages

Following are the advantages of mass production:

1. Higher rate of production with reduced cycle time.
2. Higher capacity utilisation due to line balancing.
3. Less skilled operators are required.

4. Low process inventory.
5. Manufacturing cost per unit is low.

Limitations

Following are the limitations of mass production:

1. Breakdown of one machine will stop an entire production line.
2. Line layout needs major change with the changes in the product design.
3. High investment in production facilities.
4. The cycle time is determined by the slowest operation.

Continuous Production

Production facilities are arranged as per the sequence of production operations from the first operations to the finished product. The items are made to flow through the sequence of operations through material handling devices such as conveyors, transfer devices, etc.

Characteristics

Continuous production is used under the following circumstances:

1. Dedicated plant and equipment with zero flexibility.
2. Material handling is fully automated.
3. Process follows a predetermined sequence of operations.
4. Component materials cannot be readily identified with final product.
5. Planning and scheduling is a routine action.

Advantages

Following are the advantages of continuous production:

1. Standardisation of product and process sequence.
2. Higher rate of production with reduced cycle time.
3. Higher capacity utilisation due to line balancing.
4. Manpower is not required for material handling as it is completely automatic.
5. Person with limited skills can be used on the production line.
6. Unit cost is lower due to high volume of production.

Limitations

Following are the limitations of continuous production:

1. Flexibility to accommodate and process number of products does not exist.
2. Very high investment for setting flow lines.
3. Product differentiation is limited.

1.4.2 Objectives of Production Management

The objective of the production management is 'to produce goods, services of right quality and quantity at the right time and right manufacturing cost'.

1. **Right Quality:** The quality of product is established based upon the customers needs. The right quality is not necessarily best quality. It is determined by the cost of the product and the technical characteristics as suited to the specific requirements.

2. **Right Quantity:** The manufacturing organization should produce the products in right number. If they are produced in excess of demand the capital will block up in the form of inventory and if the quantity is produced in short of demand leads to shortage of products.
3. **Right Time:** Timeliness of delivery is one of the important parameter to judge the effectiveness of production department. So the production department has to make the optimal utilization of input resources to achieve its objective.
4. **Right Manufacturing Cost:** Manufacturing costs are established before the product is actually manufactured. Hence, all attempts should be made to produce the products at pre-established cost, so as to reduce the variation between actual and the standard.

1.5 OPERATION SYSTEM

Operation system converts inputs in order to provide outputs which are required by a customer. It converts physical resources into outputs, the function of which is to satisfy customer wants, *i.e.*, to provide some utility for the customer. In some of the organization the product is a physical good (hotels) while in others it is a service (hospitals). Bus, rail, hospitals, colleges are the examples of operating system.

Ray Wild defines operation system as, "An operating system is a configuration of resources combined for the provision of goods or services."

Everett E. Adam and Ronald J. Ebert define operating system as, "An operating system (function) of an organisation is the part of an organisation that produces the organization's physical goods and services."

1.5.1 Product Vs. Service

The main differences between product and service are as follows:

(*i*) Products are visible and services are invisible.

(*ii*) There may be the time gap between the production and consumption of the product but services are produced and consumed simultaneously.

(*iii*) Products have physical dimensions and attributes, take up space in inventory, depreciate and often wear out but services cannot be stored.

(*iv*) Products may be perishable but services can never be perishable.

(*v*) Products can be evaluated against specifications and criteria but service quality is evaluated against the satisfaction of the customers.

1.5.2 Managing Operations

Managing operations can be enclosed in a frame of general management function. Operation managers are concerned with planning, organising and controlling the activities which affect human behaviour through models.

Planning

Activities that establishes a course of action and guide future decision-making is planning. The operations manager defines the objectives for the operations subsystem of the organization, and the policies, and procedures for achieving the objectives. This stage includes clarifying the role and focus of operations in the organization's overall strategy. It also involves product planning, facility designing and using the conversion process.

Organizing

Activities that establishes a structure of tasks and authority. Operation managers establish a structure of roles and the flow of information within the operations subsystem. They determine the activities required to achieve the goals and assign authority and responsibility for carrying them out.

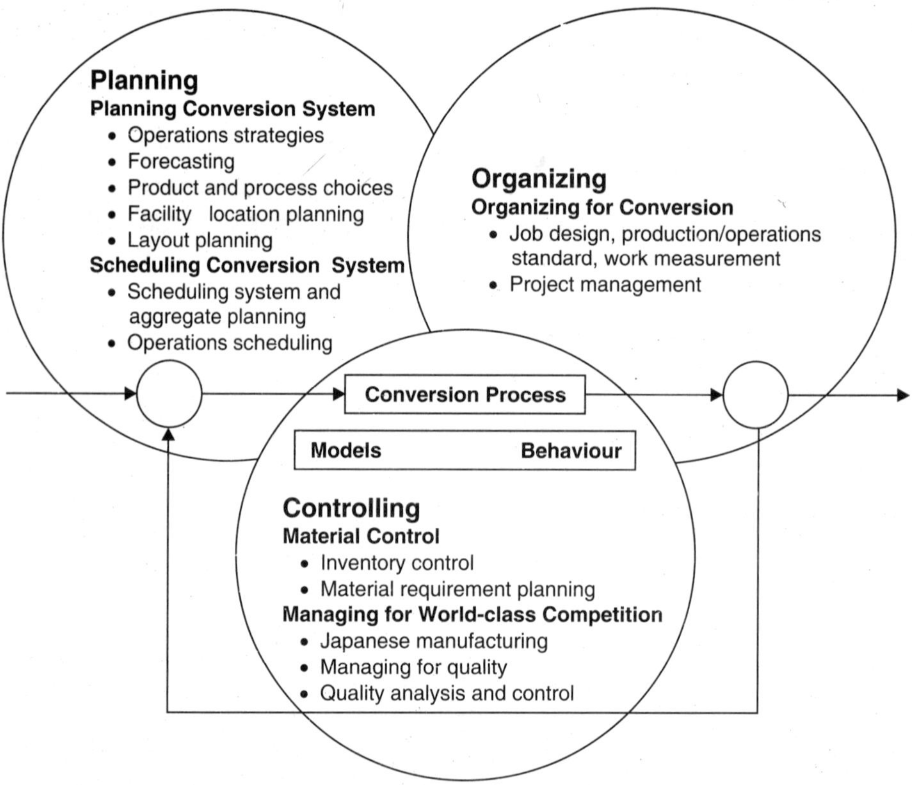

Fig. 1.2 General model for managing operations

Controlling

Activities that assure the actual performance in accordance with planned performance. To ensure that the plans for the operations subsystems are accomplished, the operations manager must exercise control by measuring actual outputs and comparing them to planned operations management. Controlling costs, quality, and schedules are the important functions here.

Behaviour

Operation managers are concerned with how their efforts to plan, organize, and control affect human behaviour. They also want to know how the behaviour of subordinates can affect management's planning, organizing, and controlling actions. Their interest lies in decision-making behaviour.

Models

As operation managers plan, organise, and control the conversion process, they encounter many problems and must make many decisions. They can simplify their difficulties using models like *aggregate planning models* for examining how best to use existing capacity in short-term, *break even analysis* to identify break even volumes, *linear programming and computer simulation* for capacity utilisation, *decision tree analysis* for long-term capacity problem of facility expansion, *simple median model* for determining best locations of facilities etc.

1.5.3 Objectives of Operations Management

Objectives of operations management can be categorised into customer service and resource utilisation.

Customer Service

The first objective of operating systems is the customer service to the satisfaction of customer wants. Therefore, customer service is a key objective of operations management. The operating system must provide something to a specification which can satisfy the customer in terms of cost and timing. Thus, primary objective can be satisfied by providing the 'right thing at a right price at the right time'.

These aspects of customer service—specification, cost and timing—are described for four functions in Table 1.2. They are the principal sources of customer satisfaction and must, therefore, be the principal dimension of the customer service objective for operations managers.

Table 1.2: Aspects of customer service

Principal function	*Principal customer wants*	
	Primary considerations	*Other considerations*
Manufacture	Goods of a given, requested or acceptable specification	*Cost, i.e.,* purchase price or cost of obtaining goods. *Timing, i.e.,* delivery delay from order or request to receipt of goods.
Transport	Management of a given, requested or acceptable specification	*Cost, i.e.,* cost of movements. *Timing, i.e.,* 1. Duration or time to move. 2. Wait or delay from requesting to its commencement.
Supply	Goods of a given, requested or acceptable specification	*Cost, i.e.,* purchase price or cost of obtaining goods. *Timing, i.e.,* delivery delay from order or request to receipt of goods.
Service	Treatment of a given, requested or acceptable specification	*Cost, i.e.,* cost of movements. *Timing, i.e.,* 1. Duration or time required for treatment. 2. Wait or delay from requesting treatment to its commencement.

Generally an organization will aim reliably and consistently to achieve certain standards and operations manager will be influential in attempting to achieve these standards. Hence, this objective will influence the operations manager's decisions to achieve the required customer service.

Resource Utilisation

Another major objective of operating systems is to utilise resources for the satisfaction of customer wants effectively, *i.e.*, customer service must be provided with the achievement of effective operations through efficient use of resources. Inefficient use of resources or inadequate customer service leads to commercial failure of an operating system.

Operations management is concerned essentially with the utilisation of resources, *i.e.*, obtaining maximum effect from resources or minimising their loss, under utilisation or waste. The extent of the utilisation of the resources' potential might be expressed in terms of the proportion of available time used or occupied, space utilisation, levels of activity, etc. Each measure indicates the extent to which the potential or capacity of such resources is utilised. This is referred as the objective of resource utilisation.

Operations management is also concerned with the achievement of both satisfactory customer service and resource utilisation. An improvement in one will often give rise to deterioration in the other. Often both cannot be maximised, and hence a satisfactory performance must be achieved on both objectives. All the activities of operations management must be tackled with these two objectives in mind, and many of the problems will be faced by operations managers because of this conflict. Hence, operations managers must attempt to balance these basic objectives.

Table 1.3 summarises the twin objectives of operations management. The type of balance established both between and within these basic objectives will be influenced by market considerations, competitions, the strengths and weaknesses of the organization, etc. Hence, the operations managers should make a contribution when these objectives are set.

Table 1.3: The twin objectives of operations management

The customer service objective	*The resource utilisation objective*
To provide agreed/adequate levels of customer service (and hence customer satisfaction) by providing goods or services with the right specification, at the right cost and at the right time.	To achieve adequate levels of resource utilisation (or productivity) *e.g.*, to achieve agreed levels of utilisation of materials, machines and labour.

1.6 MANAGING GLOBAL OPERATIONS

The term 'globalization' describes businesses' deployment of facilities and operations around the world. Globalization can be defined as a process in which geographic distance becomes a factor of diminishing importance in the establishment and maintenance of cross border economic, political and socio-cultural relations. It can also be defined as worldwide drive toward a globalized economic system dominated by supranational corporate trade and banking institutions that are not accountable to democratic processes or national governments.

There are four developments, which have spurred the trend toward globalization. These are:

1. Improved transportation and communication technologies;
2. Opened financial systems;
3. Increased demand for imports; and
4. Reduced import quotas and other trade barriers.

When a firm sets up facilities abroad, it involve some added complexities in its operation. Global markets impose new standards on quality and time. Managers should not think about domestic markets first and global markets later, rather it could be think globally and act locally. Also, they must have a good understanding of their competitors. Some other important challenges of managing multinational operations include other languages and customs, different management style, unfamiliar laws and regulations, and different costs.

Managing global operations would focus on the following key issues:

- To acquire and properly utilize the concepts related to global operations, supply chain, logistics, etc.
- To associate global historical events to key drivers in global operations from different perspectives.
- To develop criteria for conceptualization and evaluation of different global operations.
- To associate success and failure cases of global operations to political, social, economical and technological environments.
- To envision trends in global operations.
- To develop an understanding of the world vision regardless of their country of origin, residence or studies in a respectful way of perspectives of people from different races, studies, preferences, religion, politic affiliation, place of origin, etc.

1.7 FUNCTIONS/SCOPE OF PRODUCTION AND OPERATION MANAGEMENT

Production and operations management concern with the conversion of inputs into outputs, using physical resources, so as to provide the desired utilities to the customer while meeting the other organizational objectives of effectiveness, efficiency and adaptability. The following are the functions which are done by the production/operation manager.

1. *Materials:* The selection of materials for the product. Production manager must have sound knowledge of materials and their properties, so that he can select appropriate materials for his product. Research on materials to find alternatives to satisfy the changing needs of the design is his duty.
2. *Methods:* Finding the best method for the process, to search for the methods to suit the available resources, identifying the sequence of process are some of the activities of production management.

3. *Machines and Equipment:* Selection of suitable machinery for the process desired, designing the maintenance policy and design of layout of machines are taken care of by the production management department.
4. *Estimating:* To fix up the production targets and delivery dates and to keep the production costs at minimum, a thorough estimation of production times and production costs are done by production department. In competitive situation this help the management to decide what should be done in arresting the costs at desired level.
5. *Loading and Scheduling:* The production management department has to draw the time-table for various production activities, specifying when to start and when to finish the process required. It also has to draw the timings of materials movement and plan the activities of manpower. The scheduling is to be done keeping in mind the loads on hand and capacities of facilities available.
6. *Routing:* The routing consists of fixing the flow lines for various raw materials, components etc., from the stores to the packing of finished product, so that all concerned knows what exactly is happening on the shop floor.
7. *Despatching:* The production department has to prepare various documents such as job cards, route sheets, move cards, inspection cards for each and every component of the product. These are prepared in a set of five copies. These documents are to be released from production management department to give green signal for starting the production. The activities of the shop floor will follow the instructions given in these documents. Activity of releasing the documents is known as dispatching.
8. *Expediting or follow up:* Once the documents are dispatched, the management wants to know whether the activities are being carried out as per the plans or not. Expediting engineers go round the production floor along with the plans, compare than actual with the plan and feedback the progress of the work to the management. This will help the management to evaluate the plans.
9. *Inspection:* Here inspection is generally concerned with the inspection of activities during production, but the quality inspection is done by a separate quality control department, which is not under the control of production management.
10. *Evaluation:* The production department must evaluate itself and its contribution in fulfilling the corporate objectives and the departmental objectives. This is necessary for setting up the standards for future.

1.8 OPERATIONS STRATEGY

Strategy is concerned with the actions an organization takes in order to survive and prosper in its environment over the long-term. Strategy can exist at three levels in an organization: corporate, business and functional.

An organization's operations strategy comprises the totality of the actions and decisions taken within the operations function. The decisions and actions taken have a direct impact on an organization's business and corporate strategy.

An organization's operations can be a source of competitive advantage if they are managed strategically in pursuit of a clear goal for operations.There are five possible operations objectives:

1. **Cost:** The ability to produce at low cost.
2. **Quality:** The ability to produce in accordance with specification and without error.
3. **Speed:** The ability to do things quickly in response to customer demands and thereby offer short lead times between when a customer orders a product or service and when they receive it.
4. **Dependability:** The ability to deliver products and services in accordance with promises made to customers (*e.g.*, in a quotation or other published information).
5. **Flexibility:** The ability to change operations. Flexibility can comprise up to four aspects:
   (*i*) The ability to change the volume of production.
   (*ii*) The ability to change the time taken to produce.
   (*iii*) The ability to change the mix of different products or services produced.
   (*iv*) The ability to innovate and introduce new products and services.

Excelling at one or more of these operations performance objectives can enable an Organization to pursue a business strategy based on a corresponding competitive factor. However, it is important to note that the success of any particular business strategy depends not only on the ability of operations to achieve excellence in the appropriate performance objectives, but crucially on customers valuing the chosen competitive factors on which the business strategy is based. Matching operations excellence to customer requirements lies at the heart of any operations based strategy. It is unlikely that any single organization can excel simultaneously at all of the five operations performance objectives. Trying to do so is likely to lead to confusion if operations mangers pursue different objectives at different times. This lack of clarity is likely to lead to suboptimal performance and result in a failure to excel in any of the operations performance objectives. Consequently, organizations need to choose which performance objectives they will give priority to. This may result in having to trade-off less than excellent performance in one aspect of operations in order to achieve excellence in another. The concept of trade-off in operations objectives was first proposed many years ago by Skinner (1969). He argued that operations could not be 'all things to all people'. What was needed was to identify a single goal or 'task' for operations; a clear set of competitive priorities to act as the objective. The task would then act as the criterion against which all decisions and actions in operations could be judged.

Operation Strategy Process

The process of operations strategy concerns the way in which an organization develops its operations strategy. This might be top-down (*i.e.*, formed in pursuit of its business and corporate strategy), bottom-up (*i.e.*, formed from the actions and decisions taken with operations), market-led (*i.e.*, formed in response tomarket requirements) or operations-led (based on the resources and capabilities within its operations).

Operation Strategy Content

The content of operation strategy consists of the key decision areas concerned with the structure (*i.e.*, the physical attributes of facilities, capacity, process technology and supply network) and infrastructure (*i.e.*, planning and control, quality, organization, human resources, new product development and performance measurement) of operations.

EXERCISES

1. Define production and operations management.
2. Distinguish between product, service and project.
3. Why production management is known as operations management?
4. Discuss the system concept of production.
5. Discus the objectives of production management.
6. Explain the functions of production management.
7. Discuss the scope of production and operation management.

❑❑❑

CHAPTER 2

Product Design

2.1 INTRODUCTION

In the introduction to this chapter the point is made that our general image of a designer is closer to a fashion designer or a motor car designer than an operations manager. The word 'design', in its broadest sense, is right at the heart of operations management. However, many of the qualities we associate with fashion and motor car designers are also necessary for operations management 'designers'. We expect fashion designers to show flair and creativity but we also expect them to be able to understand the way fashion trends are moving. If we substitute 'the way the market is moving' for the 'way fashions trends are moving' then we can see similarities between the two types of designer. Both have to reconcile their own creative ideas (what they would like to do) with what is commercially viable (what the market wants them to do). There is no simple definition of design. But that design,

(*i*) must meet changing needs of the customers;

(*ii*) applies to products, services and processes;

(*iii*) can be managed as an operations transformation process;

(*iv*) starts with somcthing very abstract (a concept) and ends with something very specific (the final design).

2.2 PRODUCT DESIGN STAGES

The products and services move through five stages as they are designed. These stages are:

(*i*) Concept Generation

(*ii*) Screening

(*iii*) Preliminary Design

(*iv*) Prototyping and Testing

(*v*) Final Stage.

It is worth remembering however that not every product and service moves smoothly between these stages. In practice, the stages could be defined in different ways and the sequence may vary. Most importantly, there will almost certainly be recycling between the stages. So, for example, after the testing stage, it may be that the design must go right back to reconsider the original concept. In fact, at any stage the design could be recycled back to a previous stage.

2.3 CONCEPT GENERATION

New ideas may be generated by:

(*i*) Imitation

(*ii*) Adaptation

(*iii*) Invention

(*iv*) Other Sources

(*i*) **Imitation:** Imitation means making another product similar to one in the market. For example when one firm has introduced a refrigerator with automatic defrosting unit, other firm may imitate their refrigerator having similar unit. Design by imitation saves lot of research and development money. It also avoids a risk of being unsuccessful because only those designs are imitated which have proved their success in the market. It is a quick mean of developing the product.

(*ii*) **Adaptations:** Adaptation involves developing an improved product for an already existing product in the market. For example, introduction of electronic and automatic clocks against mechanically spring wound clocks.

(*iii*) **Invention:** New designs come from invention. New conceived ideas are developed and turned into new products. This procedure involves lot of money, time, risk of failure of unsuccess in the consumer market.

(*iv*) **Other Sources:** Sometimes new ideas may obtain from

(*a*) Dealers and customers.

(*b*) By advertising, asking people to send their ideas and announcing prizes for best ideas.

(*c*) Success stories of friend, relative, experience of other manufacturers.

(*d*) Chances of producing substitute of an article for which there is good demand.

(*e*) Internal sources for new ideas are employees including R & D staff, sales representatives, production workers etc. Every employee is a potential source of new product ideas and should be encouraged to put forth suggestions.

2.4 IDEA SCEREENING

The various product ideas are then carefully screened. Poor ideas are dropped and through the process of elimination only the most promising, feasible and profitable ideas are selected for further detailed investigation and evaluation.

2.5 PRELIMINARY DESIGN

The factors to be considered at preliminary product and service design stage are:

1. Marketing Aspect
2. Product Characteristics:
   (*a*) Functional Aspect
   (*b*) Operational Aspect
   (*c*) Durability and Dependability
   (*d*) Aesthetic Aspect
3. Economic Analysis:
   (*a*) The Profit Consideration
   (*b*) Standardization, Simplification and Specialization
   (*c*) Break-Even Analysis
4. Production Aspect

1. Marketing aspect

It is necessary that the proposed product will satisfy demand in the market. The demand for the product may already exist in the market and its volume can then be assessed by consumers research and sales figure for identical similar products. Market research helps to find out type of customers in the market. To have optimum quality of design for the product, customer complaints if any should be considered so that necessary modifications can be made. Demand can also be created with the introduction of a new product, either by filling in a gap in the market as by offering new properties such as novelty, appearance, or by incorporating additional features. The volume of demand is a function of multitude of factors such as consumer's liking, the availability of competitive products and market conditions etc. It is therefore essential for an enterprise to keep in touch with the market and feel its trends. Another marketing aspect of product design relates to product variety. In fact, product design is a result of some sort of compromise between infinite variety on one hand and the designer concept of ideal design on the other.

2. Product characteristics

(*i*) ***Functional Aspect:*** The Functional aspect of the product has to be carefully analysed and properly defined. The defined objective does not classify the functional scope expected. A washing machine, for example, has clearly defined objective: to wash clothing. This does not state how the washing should be carried out, whether the machine should be capable of heating the water prior to washing, whether rinsing or drying, or both are to be done by the machine and if so by what method and what should be the proportion between automatic functioning and manual supervision. A functional analysis of this kind obviously affects the design of the machine, its complexity, its shape, appearance and price. Sometimes functional aspects are detachable and usage can be left to customer's decision. For example, two-in-one cassette recorder, which can be used as a radio or tape-recorder.

There is a trend to offer functional versatility of the product, increasing the range of application and sometimes combining number of attachments to be added for a variety of functions. A food processor, for example, allows for a large number of attachments to be added for variety of functions.

Versatility of production machinery may quite often result in substantial saving in floor space and capital expenditure, and this may become one of the fundamental factors affecting design policy.

(*ii*) ***Operational Aspect:*** The product should function properly at the same time it must be easy to handle and simple to operate. The designer's problem become more critical with the trend for increased versatility. Because the increased versatility increases the complexity of product and requires certain amount of skill and intelligence to operate. The scarcity of skill is a prohibitive limitation in this respect. With the aid of different attachment, the designer should always bear in mind the time required for an operator to perform the changeover. The changeover time should be in reasonable proportion to the operation time.

(*iii*) ***Durability and Reliability:*** Durability and reliability are two factors that often determine the quality and have to be carefully considered by designer. Durability is the length of active life of product under given working conditions. Reliability is the probability of the product to perform, its intended function satisfactorily for a specified period of time under the given environment.

(*iv*) ***Aesthetic Aspect:*** Aesthetic aspect are the governing factors in the design and completely dominate it. This is especially true for many consumer goods such as automobile and household equipment as fashion goods. It is often used as means to create demand. The aesthetics are mainly concerned with molding the final shape around the basic skeleton. Many techniques can be used by designer to create aesthetic characteristics in product. These are:

(*a*) Use of special material either for the parts of housing or as additional decorations. For example, use of chromium strips, plastics, wood, glass, fabrics etc.

(*b*) Use of colour either natural colour of material concerned or by use of paints, plating, spraying. Composition and contrast of colours is of great importance to the designer.

(*c*) Texture supplements colours, either by appropriate treatment of the given surfaces as coatings.

(*d*) Line is used to break the form, also for the purpose of emphasize of parts of it, as to give a sense of continuity, graciousness and stability.

(*e*) Scaling of the product, either to a blown up size or to a small size. This creates novelty and a sense of completeness.

(*f*) Shape can be exploited to accentunate particular feature, to create a sense of spaciousness, richness.

(*g*) Packaging, especially for small items. Novelty and attractiveness of packaging is often transferred in the minds of customer.

3. Economic Analysis

(*a*) Economic analysis is the key to management decision in product design. It includes evaluation of capital expenditure required to manufacture the new products, the estimation of product cost per piece and the reasonable margin of profit. If the organization wants to increase its profit the following methods can be used:

(*a*) Increase the sales price

(*b*) Increase the market

(*c*) Reduce total cost

If both profit per piece and size of market increase, a substantial improvement in total profit will be achieved. This can be done by continous search for better methods, better processes, better and cheaper alternative materials and their utilization and better management to reduce overheads. Reducing production cost and thereby expanding the market, while sustaining accepted quality.

(*b*) The three S's Standardization, Simplification and Specialization are at the root of any economic analysis of product design development.

1. Standardization: It is a process of defining and applying the conditions necessary to ensure that a given range of requirements can normally be met with a minimum variety and in a reproducible and economic manner on the basis of best current technique. It is applicable to all factors of production namely men, machines, methods and finished goods. These standards can become the basis to evaluate the performance of various components of production in a manufacturing process. Behel, Smith and Stackman defines standard as "a criteria of measurement, quality, performance, practice established by custom, consent or authority and used as a basis for comparison over a period of time."

Standardization covers a wide field of activity:

— Physical dimensions and tolerances of components within a defined range.

— Rating of machines or equipment (in units of energy, temperature, speed etc.)

— Specification of physical and chemical properties of materials.

— Methods of installation to comply with minimum precautionary measures and convenience of use.

Advantages of Standardization

1. Standardization in designing, purchasing of raw materials, semi-finished and finished goods and of the manufacturing process tries to eliminate wastage and reduces the cost of production.
2. Reduction in variety of raw materials means reduced investments in stocks and reduced inventories.
3. By minimising and standardizing the operations in production process it provides facility to introduce mechanisation and use of more specialized tools and equipments.
4. Production in larger quantities can be planned which results in less set up costs.
5. Less skill is required in manufacturing and assembly.
6. Reduction in repair and maintenance cost and marketing expenses.
7. Reduction in book-keeping and other paper work.
8. Better inspection and quality control is possible.
9. Because of large purchase quantities involved, the economies in buying and selling can be availed.

Disadvantages of Standardization

1. Change in public taste or liking may seriously affect a company producing only standardized product range.
2. Too much standardization has an adverse effect on the efficiency and morale of the workers. They in the long run feel bored and fed up in doing the same routine again and again.
3. Due to monotonous work the spirit of challenge and initiative vanishes with the passage of time.
4. It becomes difficult to introduce new models because of less flexible production facilities and due to the high cost of specialized production equipment.
5. Standards once set, resist change and thus standardization may become obstacle to progress.
6. For small scale enterprises standardizations may not be advantageous.

Preferred Number

According to American Standards Association, "preferred numbers" are defined as "series of numbers selected to be used for standardization purposes in preference to other numbers. Their use leads to simplified practice and they should, therefore, be employed whenever possible for individual standard sizes and ratings, or for a series thereof," in applications relating to important or characteristic linear dimensions (such as diameters and lengths) or specifications of areas, volumes, weights and capacities, rating of machinery and apparatus.

The preferred numbers were first introduced by Renard in 1870, and therefore the series is sometimes referred as the Renard series.

Suppose a manufacturer wants to produce containers having between 10 and 100 liters capacity. In selecting the series he can choose a 5 series, *i.e.*, covering the range of 10 to 100 in 5 steps or a 10-series, *i.e.*, covering the range of 10 to 100 in 10 steps or a 20-series, *i.e.*, covering the range of 10 to 100 in 20 steps or a 40 series, *i.e.*, covering the range of 10 to 100 in 40 steps.

If the said manufacturer decides to adopt the 5-series, his capacity rating to the geometric progression series will be:

First size 10 (given)

$$\left.\begin{array}{l}\text{Second size } 10 \times q \\ \text{Third size } 10 \times q^2 \\ \text{Fourth size } 10 \times q^3 \\ \text{Fifth size } 10 \times q^4\end{array}\right\} \text{5 steps}$$

Sixth size $10 \times q^5 = 100$ given

Hence, $q^5 = 10$

or $q = 10^{1/5} = 1.585$

The calculated sizes would be

First size 10.00	*i.e.,* select 10	5 steps
Second size 15.85	*i.e.,* select 16	
Third size 25.12	*i.e.,* select 25	
Fourth size 39.82	*i.e.,* select 40	
Fifth size 63,11	*i.e.,* select 63	
Sixth size 100.00	*i.e.,* select 100	

The basic preferred numbers for all those series, as suggested by the international system are shown in Table 2.1.

Table 2.1: Basic Preferred Numbers (10 to 100)

A	B	C	D	E	F	G	H
10	10.6	11.2	11.8	12.5	13.2	14.0	15.0
16	17.0	18.0	19.0	20.0	21.2	22.4	23.6
25	26.5	28.0	30.0	31.5	33.5	35.5	37.5
40	42.5	45.0	47.5	50.0	53.0	56.0	60.0
63	67.0	71.0	75.0	80.0	85.0	90.0	95.0

Where the given rounded numbers do not depart from the theoretical calculation by more than 1.3 percent. The 5-series is as given in the column A, to obtain the 10-series column to E should be added, hence the series would be: 10, 12.5, 16, 20, 25, 31.5, 40, 50, 63 and 80. Similarly, for the 20-series would be column A, C, E and G (*i.e.,* 10, 11.2, 12.5, 14, 16, 18 etc.). And for 40 series would be from column A, B, C, D, E, F, G and H, *i.e.,* 5 series = $(20)^{1/5}$ =1.5849 or 60% steps column A.

10 series = $(10)^{1/10}$ = 1.2589 or 25% steps column A, E

20 series = $(10)^{1/20}$= 1.1220 or 12% steps A, C, E, G

and 40 series = $(10)^{1/40}$= 1.0593 or 6% steps, A, B, C, D, E, F, G, H

2. Simplification: *W.R. Spreigal* and *R.H. Lansburg* define, "simplification as the elimination of super fluous varieties, sizes, dimensions etc. Simplification makes the product, assembly or design simpler, less complex or less difficult. Simplification should be done before standardization. Simplification is a constant source of disagreement between the sales department and the production personnel. A production engineer prefers little variety, minimum set-ups and long runs. Simplification enables the production department to improve planning, achieve higher rates of production and machine utilization and simplify control procedure. The marketing (sales) department on the other hand, strives to satisfy the customer by giving him a choice and by offering him the closest to his choice.

Advantages of Simplification

1. Simplification reduces inventories and thus results in better inventory control.
2. Due to simplification in operations the efficiency of the production process increases and this leads to more productivity.
3. Simplification provides quick delivery and better after sales service.

4. Simplification implies fewer parts and fewer the parts, the lower the production costs.
5. It simplifies planning and production methods.
6. Simplifies inspection and control.
7. Reduces sales price (through production simplification and reduction of distribution costs), hence increase in sales.
8. Facilitates repair service and maintenance of the product.
9. Reduction of storage space.
10. Less overheads and handing expenditures.

3. Specialization: Specialization implies expertise in some particular area or field. A worker is said to be specialized in a work when he acquires skill and proficiency in it by concentrating solely on it. A factory producing only separate plug is a specialist in its production. Another example may be tyre industry, ball bearing manufacture etc. In case the company cannot fix the extent of specialization in product line, the enterprise is likely to grow to unmanageable and unprofitable proportions. The optimal degree of specialization is such an important factor in the welfare of the company. When an enterprise thinks of introducing a new product, the management has to decide whether the new product will suit their facilities available, without diluting the skills or unduly increasing the complexities of business.

Advantages of Specialization

(*i*) Increase in output and reduction in per unit cost of production.
(*ii*) Higher productivity.
(*iii*) Improvement in the quality of finished product.
(*iv*) The firm can afford to carry out R&D for further improvement.

Disadvantages

(*i*) Less flexibility in adjustment to changed situation.
(*ii*) Monotony and boredom may adversely affect the efficiency.

(*c*) **Break-Even Analysis:** The management has to decide that what is the minimum quantity that must be produced and sold so as to run the business profitably. The quantity can be bound by applying break-even analysis. To understand the break-even analysis one must understand the cost structure of a product. The total cost of the product is made up of

(*a*) Fixed costs (*b*) Variable cost

(*a*) **Fixed Cost:** Fixed cost consists of all those costs or expenditure which may not vary with the quantity produced. For example, rent, interest, factory power costs, salaries of managers etc.

(*b*) **Variable Cost:** This consists of those costs, which will vary directly with the quantity produced. For example, are direct labour cost, material costs and fuel consumption charges of machines, when we add overhead charges to sum of fixed cost and variable cost, we get the total cost.

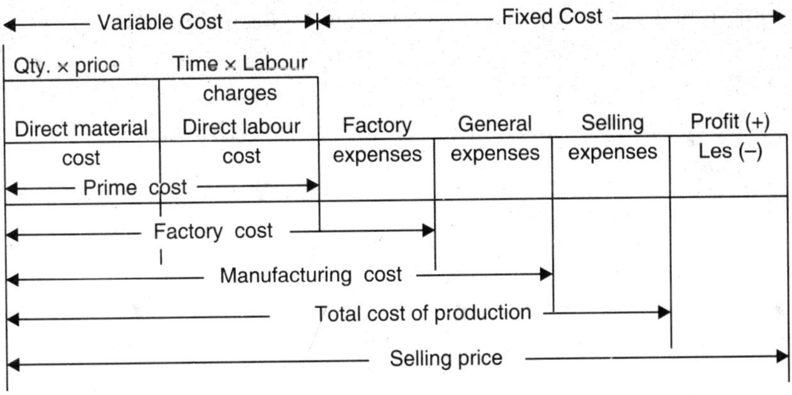

BEP makes use of fixed and variable cost to indicate the range of value necessary for profitable operation of the plant. The part of revenue, which is absorbed by fixed cost, is known as contribution. Contribution is equal to difference between Sales Revenue and Variable Costs.

Contribution (C) = Sales Revenue (S) – variable cost (V)

If F = fixed cost and P = Profit in rupees, then

$$S = F + V + P$$

or

$$P = S - F - V$$
$$= (S - V) - F$$
$$= C - F$$

If C is less than F then we can say organization is running in loss.

Now consider that

V = variable cost per unit in Rupees.

S = Sales price per unit in Rupees.

F = Fixed costs in Rupees.

Q = Break-even point.

At Break-even point the Revenue earned by the company is equals to the expenditure incurred.

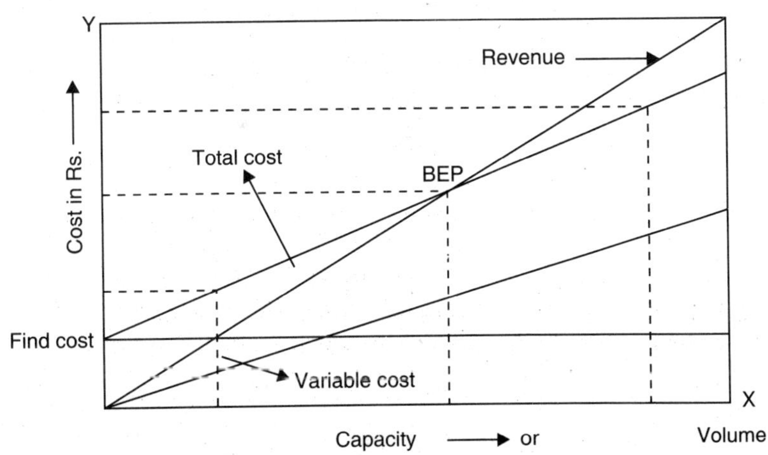

Fig. 2.1 Break-even chart.

Therefore $F + Q \times V = Q \times S$

or Q in Units $\dfrac{F}{(S - V)} = \dfrac{\text{Fixed Costs}}{\text{Contribution per unit}}$

Figure 2.1 gives the graph showing break-even point.

Break-even analysis is a part of marginal costing, basically it is concerned with finding the point at which Revenues and costs agree exactly, hence the term, break-even. The Break-even point is, therefore, the volume of output at which neither profit is made nor a loss is incurred.

As a result of calculating break-even point the financial department can realise that many valuable facts, all essential for operating a business enterprise, can emerge from the exercise. Some typical problems that can be solved from break-even analysis are:

(*i*) A reasonable return on capital employed.

(*ii*) To compare a number of companies by arranging probable earnings from each in order of magnitude.

(*iii*) Determination of the price of a product, which will give the desired break-even point and profit.

(*iv*) Calculation of costs and revenues for all possible volumes of output' thereby helping in fixing sales budget.

(*v*) Variable costs per unit can be calculated quite easily from a break-even chart.

(*vi*) The cash involved in obtaining on particular volume of output.

Example 2.1: *The fixed costs per year ₹40,000. Variable cost per unit for the single product is ₹2. Estimated sales for the period is ₹1,00,000. The number of units involved coincides with the expected volume of output. Each unit is sold at ₹10 each. Find the break-even point.*

Solution: Break-even point (BEP) = $\dfrac{F}{1 - \dfrac{V}{P}}$

where F = Fixed costs in Rupees.

V = Variable cost in rupees per unit.

P = Selling price per unit in rupees.

$$BEP = \frac{40,000}{1 - \dfrac{2}{10}} = ₹\ 50,000$$

= Total sales value (S) in Rupees.

Break-even point in units = $\dfrac{F}{P - V}$ Units

where F = Fixed costs in Rupees.

P = Selling Price

V = Variable cost per unit in Rupees.

$$= \frac{40,000}{10-2} = 5,000 \text{ units}$$

Example 2.2: *Variable costs are 60% of the sales and fixed costs are ₹ 30,00,000. Compute profit, if total sales are ₹ 10,000,000.*

Solution: Since $C = S - V$,

where C = Contribution in Rupees

S = Sales Revenue in Rupees

V = Variable cost in Rupees

$C = S - 60\%$ of S $\qquad C = 40\%$ of S

$$\therefore \quad C = \frac{40}{100} \times 10,000,000 = 40,000,000$$

$$P = S - V - F = C - F = 4,000,000 - 3,000,000.$$

$$= ₹\ 1,000,000$$

4. Production Aspect: Last but not least in most of factors influencing design is the production aspect. The practical production problems should also be considered in the design. "Design for production" has become a motto among designers.

The following aspects of production engineering are considered:

(*a*) Selection of processes that will be most suitable and economical for the purpose considering:

- (*i*) Volume of production.
- (*ii*) Utilization of existing equipment.
- (*iii*) Selection of signs and fixtures and other production aids the use of which may affect the design of components.
- (*iv*) Sequence of operations and methods.
- (*v*) Limitation of skill.
- (*vi*) Application of new production processes.

(*b*) Utilization of materials and components with the view of:

- (*i*) Selection of materials having appropriate specification.
- (*ii*) Selection of method or design to reduce waste and scrap.
- (*iii*) Using standard components and assemblies.
- (*iv*) Having interchangeability of components and assemblies within the product.

(*c*) Selection of appropriate workmanship and tolerances that satisfy quality requirements, but which are at same time compatible with the precision and quality that can be attained through the available processes. Specification of quality may also affect the selection of processes.

2.6 REQUIREMENTS OF GOOD DESIGN

1. Customer's satisfaction
2. An adequate profit

In order to achieve customer's satisfaction:

(*i*) The product should meet functional requirements.

(*ii*) It should be of proper quality so as to achieve desired degree of accuracy and reliability.

(*iii*) It should be easy to operate.

(*iv*) It should be easy to repair and service.

(*v*) It should have good appearance.

(*vi*) It should be durable.

(*vii*) It should have reasonable price to compete with other equivalent products in the market,

In orders to get adequate profit

(*i*) It should be easy to manufacture the product within the available resources.

(*ii*) It should have minimum number of parts.

(*iii*) It should adopt latest technology so that manufacturing requires minimum cost per unit of production.

(*iv*) Standard component parts should be used whereever possible.

(*v*) Manufacturing process should be decided on the basis of volume of production.

(*vi*) A well designed product should be easy to pack and distribute.

2.7 PRODUCT LIFE CYCLE (PLC)

Product life cycle is a pattern of demand of a product throughout the life of a product. Product go through a series of stages, beginning with low demand during market development, proceeding through growth, maturity and then finally decline. The product after being developed is introduced or launched into the market and there is no pressure to earn immediate profit. The main aim of this stage is to create awareness about the product among the customers. The organization decides price for the product at which it is to be launched in the market.

After introduction phase product enters into growth stage. Products become more profitable. More competitors enter into market that offer similar quality at cheap prices. There is an increase in advertisement cost and focus is on building brand. The share of company in market tends to be stable.

After growth stage product enters into mature stage. Here sale of product stabilises and grow at a decreasing rate. Producers try to differentiate the product from the competitors product. At this point market reaches a saturation point. Price wars and intense competition occur in the market. The producers leave the market due to less profit margin.

After maturity stage there is downfall in the market. There is intense piece cutting by the producers in the market and many producers withdraw from market.

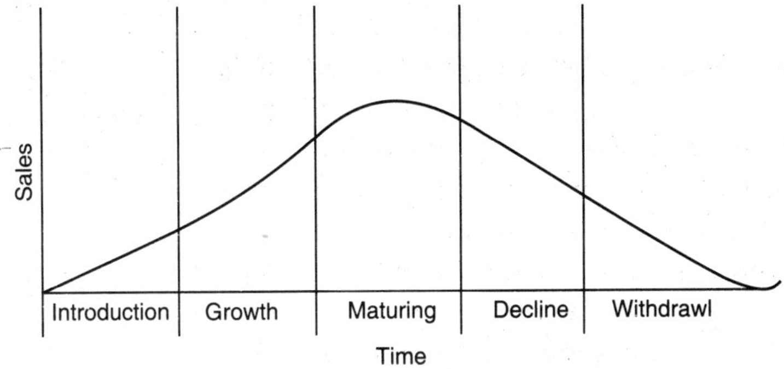

Fig. 2.2 Stage of Product Development.

2.8 MINIATURISATION

Moore says, if your product is of room size then make it desk size. If it is of desk size then make it portable. For industry in general we can say make it smaller and lighter in weight. But there is always some limit to it. Everything cannot be miniaturized. Beds cannot be smaller than the size of a person. Sometime the whole product may not be miniaturised but some parts can be reduced in size. The main characteristics of miniaturisation are:

1. It results in using less material.
2. There is saving in space and reduction in weight.
3. It is easy to handle and transport.

2.9 MODULAR DESIGN

In modern times the customer wants a great variety of products to choose the one he likes and affords best. Therefore, the manufacturers are forced to produce a variety of products. This effects the product design function and management is interested in finding ways in which combination of product can be designed and produced at a minimum cost. The use of standard sub-assemblies give rise to numerous end products in different combinations. This concept is known as modularity. The products are designed to be made of detachable component as sub-assemblies so that whenever any part or component fails it can be replaced by new one maintaing stock of modules and in assembling them promptly enables the manufactures to put forth diversity in customer's hand. For example, in selecting a personal computer the modules may consist of video monitors, key boards and printers.

Advantages

(*i*) Diversity can be offered to customers.

(*ii*) The use of detachable components enables to replace the worn out component by a new one.

(*iii*) Production efficiency increases as modules are standardized.

(*iv*) Material planning and inventory control can be simplified since standard modules are used.

(*v*) Adaptation to market segment is possible.

(*vi*) Fewer standard modules facilitates tighter quality control and standard testing procedure.

2.10 COMPUTER AIDED DESIGN (CAD)

Computer aided design uses computational and graphics software and has substantially enhanced design productivity. The geometry of the component can be graphically displayed and manipulated easily on video monitors. Alternative designs can be posed and evaluated more quickly.

Computer aided design may be defined as type of design and drafting activity which makes use of computer to develop, analyse or modify and engineering design. Modern CAD systems are based on Interactive Computer Graphics (ICG). Interactive computer graphics denotes a user-oriented system in which the computer is employed to create, transform, and display data in the form of pictures or symbols. In ICG system, the monitor is the interface between the computer and the user. This interface sets up a two-way (interactive) communication.

The computer upon receiving the input signals, can modify the displayed picture. To the user, it appears that the picture is changing instantaneously in response to his commands. He can give a series of commands, each one generating a graphical response from the computer. In this way he maintains a "conversion" or "dialogue" with the computer. Since a human is required to operate the system, CAD is also referred as 'Interactive Computer Graphics'.

The following example illustrates the application of CAD:

An engineer designing a new car body can see the drawing on the screen. He can make it turn slowly, thus providing different views. He can quickly make changes which will be instantaneously shown on the screen. The changes will also get recorded in the storage. He can reduce or enlarge any part of the drawing and can show the various components in different colours. He can test the performance of the design through a simultation exercise. Overall the design cycle is reduced drastically and the new car can be on the road in a remarkably short time.

Advantages of CAD

1. CAD systems are "user friendly". The combination of human *brain* and the machine produces better results in much less time. It thus increases the productivity of the designer. This productivity improvement results in lower design cost as well as shorter design completion time.
2. The computer works very accurately, leading to a better quality drawings.
3. The drawings can be stored in "database" (Database is a collection of data, viz., design details, graphics etc., in a single location for future use by different programmers for variety of applications).
4. With the ability to interact with the computer, you can quickly correct a design/drawing error and see a revised picture on the screen.
5. CAD is used for creation modification, analysis and optimization of designs for improved engineering productivity. An engineer can try out different idea by just sitting at the terminal without making any prototype.

6. Visual modelling of any object/engineering component is possible.
7. "Colour graphics" help to display more distinct informations on the screen, highlight certain important feature etc.
8. It improves communications by providing better engineering drawing, more standardization in the drawing, better documentation of the design, fewer drawing errors, and greater legibility.

2.11 COMPUTER AIDED MANUFACTURING (CAM)

Computer-aided manufacturing can be defined as the use of computer systems to plan, manage and control the operations of a manufacturing plant through either direct or indirect computer interface with the plant and production resources. The applications of computer-aided manufacturing fall into two broad categories:

(*i*) **Computer monitoring and control:** These are the direct applications in which the computer is connected directly to the manufacturing process for the purpose of monitoring or controlling the process.

(*ii*) **Manufacturing support applications:** These are the indirect applications in which the computer is used in support of production operations in the plant, but there is no direct interface between the computer and the manufacturing process.

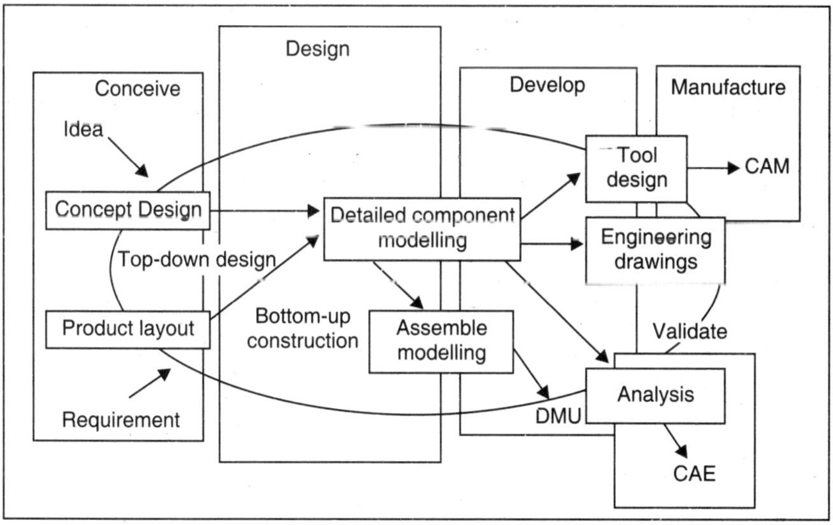

Fig. 2.3 CAD CAM Working Process.

- CAE Computer Aided Engineering used for testing the model whether its working properly or not.
- DMU Decision Making Unit is often referred to as a group of people involved in the process of making a decision (*e.g.*, purchasing a product, service)

2.12 PROTOTYPING AND TESTING

At this stage after the product is designed it is converted into a prototype development, giving a visual image of the product. This model is subjected to test to see whether it function properly or not and if the design can be corrected or improved. It is important to evaluate the work constantly during the project.

2.13 FINAL STAGE

After testing the design is finalised. The next step is branding, packaging and labelling the product. Marketing programme development is then done for introducing the product. A description of the target market size and characteristics is done and expected sales, market share and profits for the first few years is worked out on the basis of planned pricing, distribution and promotion strategy for the product.

2.13.1 Test Marketing

Once the product has been developed, it has to be tested in the actual market place to learn how consumers react to the product's features, its performance and its marketing programme including branding, packaging, price and distribution. Test marketing decisions include deciding geographical areas where testing is to be done and deciding the strategies like packaging, advertising and sampling. The product is manufactured on a limited scale and introduced in a well selected test market before starting a full scale production. Customer's reactions are then studied under normal market conditions. The reactions of the customer will give additional valuable clues for the product development and necessary modifications so as to make it more acceptable to the customers.

2.13.2 Commercialisation

Once the test marketing gives a green signal the company can undertake the full scale manufacture, considering the necessary modifications. The marketing management can then launch a full fledged advertising and promotion campaign for distribution of the product in the entire market.

Marketing management will have to undertake constant checking of the new product throughout its life cycle. Product improvement search will be a continuous affair to introduce the necessary improvements, modifications, innovation etc., in the existing product on the basis of changing consumer's preferences as well as on account of technological developments.

In general, the following points should be considered while selecting and developing a product:

1. Existing demand for the product in local market.
2. Continuity of demand.
3. Competitors for the same product, their share of market.
4. Market potential, *i.e.,* scope for future demand.
5. Scope for export market.

6. Whether the product will serve as an import substitute.
7. Degree of profitability.
8. Government policies, concession, liberal taxation policies, incentives, facilities available.
9. Ancillary product for the existing parent industries.
10. Availability of technical know how, raw materials, machinery and required skill.

It is advisable to gather information on the above points from different channels like commerce periodicals and trade journals brought out by various Government and Private institutions, material published by large, medium and public sector enterprises etc.

2.13.3 Consumer Adoption Process

A new product has to survive beyond its introductory stage and consumers have to not only make trial purchases but have to sustain their buying through repeat purchases and ultimately become regular users of the new product. Consumers typically go through a series of stages in reaching the regular user stage.

This process is termed as adoption process:

Awareness → Interest → Evaluation → Trial → Adoption.

At the 'awareness stage' the consumer knows about the existence of the product but lacks information about uses, benefits, features and so on.

At the 'interest stage' the customer considers whether the product would be useful.

At the 'trial stage' the customer makes one or two trial purchases to test out the product's usefulness.

At the 'adoption stage' the customer decides to become regular user of the new product.

The adoption rate is not the same for all products. Some products may be quickly adopted while some may take a longer time.

Following product characteristics affect the rate of adoption:

(*a*) **Relative advantage:** The greater a product perceived superiority over previous products, the faster is the adoption rate.

(*b*) **Compatibility:** It is the extent to which a product is consistent with customer's cultural values and the perceived personal and social risks associated with using the product.

(*c*) **Complexity:** It is the degree to which a new product is difficult to understand or use.

(*d*) **Trialability:** It is the degree to which the product can be used on a limited basis. Free samples, small sized introductory packages, money back guarantees, test turns and incentives to encourage trial are used by the marketer to help customers try their product.

2.14 PROCESS DESIGN

Process design can be defined as a set of activities performed to modify and add value of the input and convert it into the desired output. The process design is interrelated to the production design.

A process design requires effective combination of various sub-activities working together in coordination with each other. For example; while manufacturing parts of motor bikes design process involves many sub-activities such as procuring materials, running production process, keeping quality checks, maintenance of facilities and so on. All the sub-activities work in coordinated manner to make an effective process design.

You can adopt any of the following design processes depending on the requirement and constraints of a product:

- Assembly line
- Continuous flow production
- Batch production
- Job shop production

During the life cycle of a process design the cost of manufacturing a unit of a product diminish over a period of time. From the product start-up to the decline there are changes in the manufacturing process, volume and rates of process innovation. The process designing typically job shop at the start-up and moves towards a continuous flow technology. Volume and automation is low during the start-up and high at the time of decline. Figure 2.4 shows stages of process design lifecycle.

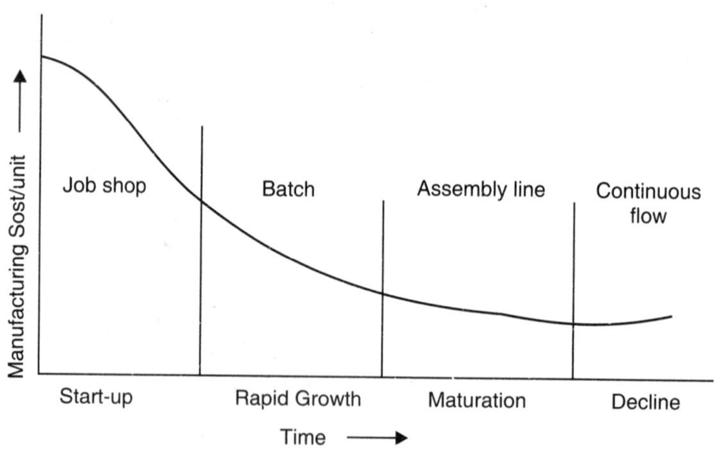

Fig. 2.4 Stages of process design lifecycle.

2.15 INTER-RELATIONSHIP BETWEEN PRODUCTION DESIGN AND PROCESS DESIGN

Following are the ways in which production design and process design are interrelated each other:

Basic Functionalities: Design of a product is based on the features of a product and functions it going to perform. Process for designing a product is planned keeping in mind these features.

Cost of manufacture: Manufacturer set the design of a product keeping in mind various costs related to manufacturing of the product, such as cost of raw material, advertisement cost and carriage inward etc. You plan the process for designing considering the planned cost. The design process should not exceed the total planned cost to design a product.

Specification: There are certain specifications to be followed to design a product, such as size and type of the product. Design process should keep in mind these specifications.

Quality and Volume: Product is designed considering the quality standards and quantity set for a product. The process for designing a product is set keeping in mind the quality of the product to be manufactured.

EXERCISES

1. Explain the concept of product design and development.
2. Explain the need for product design.
3. What are requirements of a good design?
4. Define Product Life Cycle. Describe various stages in PLC.
5. Explain the product development techniques.
6. Define computer aided design. What are advantage of CAD?
7. Define modular design and miniaturization.
8. What is a process design? Bringing out the steps involved in designing a process.
9. Discuss the inter-relationship between process design and product design.

❑❑❑

CHAPTER 3

Capacity Planning

3.1 INTRODUCTION

Capacity is the maximum possible output of a system under normal design or planned conditions in a given time period. Capacity means the ability to carry out stated objectives while the term 'capacity building' refers to a process that improves the ability of a person, group, organization, or system to meet the objectives. Capacity can be measued in terms of output or input. For a manufacturing firm the capacity is measured in terms of output. For example, a company manufacturing a single product X, and its resources can produce 1000 units per day, we can quote the capacity of plant is to produce 1000 units of product per day. In case of multiproduct companies it is difficult to mention in terms of products, as all the products share the same productive resources. In such cases, the capacity is stated in aggregate units per time period. For a service firm where output measures are difficult the capacity is measured in terms of input. For example, a service station can state its capacity as 10 four wheelers per day. A transport company can state its capacity as 60 seats per trip. Capacity utilization relates actual input/output. Capacity utilization expressed as a percentage and is given by the ratio: Actual capacity used to Design capacity.

3.2 TYPES OF CAPACITY

The capacity of a production unit/service industry is its ability to produce or do that which the customer requires. In production and operations management, three types of capacity are often referred to:

(*i*) **Potential Capacity:** The capacity that can be made available to influence the planning of senior management (example is helping them the make decision about overall business growth investment etc). This is essentially a long-term decision that does not influence day-to-day production management.

(*ii*) **Immediate Capacity:** The amount of production capacity that can be made available in the short-term. This is the maximum potential capacity that is used productively.

(*iii*) **Effective Capacity:** Not all productive capacity is actually use or usable. It is important for production managers to understand what capacity is actually achievable.

3.3 CAPACITY PLANNING

When we talk of capacity, we sometimes refer to maximum and minimum capacity. Maximum capacity concentrates on how best to utilize the available resources to produce the maximum required and the minimum capacity indicates that to produce below this capacity is not economical to the unit, as it has to meet the break-even point. Capacity planning is concerned with defining the long term and short term capacity needs of organization and determining how those need will be satisfied. Capacity decisions consider consumer demands and the available resources and co-ordinate both to decide to estimate the required capacity.

3.4 FACTORS AFFECTING CAPACITY PLANNING

Following are the factors that influence the capacity of an organization:

- *Product and Service Factors:* The capacity of an organization depends on the type of product and the type of services that the product provides. For example, the capacity; in terms of output will be very high if the organization manufactures automobile spare parts. On the other hand, if the organization manufactures products such as trucks or cars, then the capacity will be low.
- *Process:* The manufacturing process is a key factor that affects capacity planning. For example, capacity planning for a manufacturing unit in which most of the tasks are done manually will be different from the unit in which advanced machinery is used for production.
- *Availabilities of Facilities:* Availability of facilities such as use of advanced technology and improved means of communication are the factors that affect the capacity of an organization infrastructure facilities such as power supply and limitation of space can affect the capacity of a production unit.
- *Human Factors:* Human factors such as workers also affect the capacity planning of a company. For example, skilled workers can produce a better quality of goods than unskilled and untrained workers.
- *Supply Factors:* Supply of the material also affects the capacity of the company. Timely and assured supply of material improves the ability to produce goods. For example, an organization has to import raw materials from far of place. Delay in getting raw materials due to lack of proper transportation facilities affect the capacity planning.
- *External Factors:* External Factors such as investors and government policies affect the capacity planning. The amount invested by the investors affect the capacity planning of an organization. Capacity is also affected by the rules such as, maximum limit of production and tax limits as set by the government for an organization.

3.5 CAPACITY PLANNING PROCESS

The capacity planning process involves the following activities:

1. Assessing the Existing Capacity

Different organization use different ways to measure their production capacity. For example, General Motors Corporation defines capacity in terms of automobiles produced per year. Another organization uses tons of cheese produced per year to define capacity. However, it is very difficult to define the capacity in terms of output for the service organizations. For such type of organizations capacity is measured in terms of input. For example, in case of a law firm, capacity can be expressed in terms of the number of attorneys employed per year. So, we can conclude that the capacity of an organization is expressed in terms of the inputs or outputs per unit time. Sometimes, it is not possible to measure the capacity of an organization accurately due to variations in the day-to-day schedule. For example, in a car-manufacturing organization, workers may be absent or machines may break down. In such a scenario, you need to schedule variations and repair the machines. These variations occur frequently in an organization. Therefore, it is very difficult to define the capacity in precise terms.

2. Forecasting Future Capacity Requirements

To estimate the future capacity requirements of an organization, you should keep in mind the short-term and long-term requirements of an organization.

- *Short-term requirement:* These can be determined by forecasting the demand of the product or services over a period of one year. After estimating the short-term requirement, manager compares these requirements with the existing capacity and adjusts the capacity of the organization according to requirements.
- *Long-term requirement:* These can be determined based on the marketing plan, product development and life cycle of a product. Due to change in technologies and market pattern, it is very difficult to forecast the future demand of products for ten to fifteen years. Moreover it is also not sure that the organization will be producing the same products in future. It may be possible that in future an organization produces the same products, but the method of the production may change. Hence, it is important that you forecast the change in technology and the product demand to determine the long-term capacity requirements.

3. Identification of Alternative Ways to Modify Capacity

After determining the existing and future capacity requirement, you may also need to identify the ways to modify the existing capacity of an organization. The categorization for modifying capacity are:

(*i*) ***Long-term capacity decision:*** With the economy expanding day-by-day, you need to expand your business to meet the increasing customer demand. In such a scenario, you need to permanently increase the capacity of your organization. However, along with expanding the capacity, you also need to consider the approaches used for contracting the capacity. Following are the two methods to modify the capacity in long run:

(*a*) *Long-term capacity expansion:* The firm can expand capacity by the following ways:

(*i*) The firm may give a sub-contract to some other company. The sub-contractor may produce the good for the main contractor.

(*ii*) The firm may develop new sites, construct buildings or buy equipment for the purpose of expanding its capacity.

(*iii*) The firm may update on modify its existing facilities in order to increase its capacity to produce.

(*iv*) The firm may decide to acquire other companies, facilities as the resources.

(*v*) The firm may decide to restart its facilities that may be put on hold previously.

(*b*) *Long-term capacity reduction:* The firm can reduce its capacity by the following way:

(*i*) Capacity reduction may happen by way of selling off the existing resources and by laying-off the employees.

(*ii*) The firm can temporarily shut down the facilities. This has to be accompanied by the transfer of employes to some other concern under the same management or some other concerned firm.

The facility can be reduced by way of developing and phasing in new products also. This decision is accompanied by the decision to phase out the old products.

(*ii*) ***Short-term Capacity Decisions:*** Short-term capacity planning is done for a period consisting of less than a month. It deals with fluctuations in demand due to seasonal and economic factors and to match the available capacity with demand as adjusting the demand to the available capacity. Manufacturing firms with sufficient equipment capacity may use overtime as a second and third shift to meet a temporary raise in demand. This may also be satisfied by routing the work to alternate processes as subcontracting a part of order to the outsiders as sometimes the firms accumulate the inventory by stocking the products manufactures during the period when the demand is lean. Service industries tend to rely more heavily on work scheduling and strategies aimed at shifting demand to suit their capacity, such as using flexible work hours, hiring part-time employees and overtime work scheduling to meet peak demand. Professional service, such as medical, legal offices use an appointment system to smooth out demand. Capital service systems, such as transport system as airway accomplish the same thing by adopting a fixed service schedule. This arrangement forces the customers to adopt to the system is capacity. Service unit such as repair shops simply use queues and delay the delivery of the service until they can produce it with their regular workforce.

4. Evaluation of Capacity Requirement

After modifying the existing capacity of an organization, you can evaluate the short-term and long-term capacity requirements with the help of following modes.

- *Linear programming:* This model follows a mathematical approach to evaluate short-term capacity requirements. This approach helps maximize the profits of an organization by selecting the optimal combination of resources. This model is used to identify the difficulties in measuring capacity in a multiproduct firm. For example, a company produces two product, A and B, the product A needs 6 machine hours and 5 kg of raw materials and the product B need 3 machine hours and 4 kg of raw materials to produce 1 unit of A and B. Assume that the raw materials and machine hours are limited. The maximum availability of raw materials is 240 kg/month and machinery is 300 hours/month. The profit margin for the product A is ₹ 8 and for the product B is ₹ 9 per unit. The linear programming model helps determine the number of units of product A and B that the company should produce to optimally utilize resources for achieving maximum profits.
- *Decision tree analysis model:* This model follows the decision tree analysis approach to evaluate long-term capacity requirement. In the decision tree analysis model, decision tree is diagrammatically represented so that all the decision points to identify alternative decisions for capacity expansion can be shown. This model perform the following activities:
  (*i*) Identify the order in which all the decision points occur.
  (*ii*) Identify the alternative decision for each decision point.
  (*iii*) Identify several different outcomes that occur after each decision.
  (*iv*) Construct a tree diagram to represent various decisions and their outcomes.
  (*v*) Estimate the probability of each outcome.
  (*vi*) Estimate the profit of the outcome of each decision.
  (*vii*) Evaluate the expected value of each decision.
  (*viii*) Select the appropriate decision.

5. Selecting the Suitable Alternative

After evaluating the short-term and long-term capacity requirements, a company selects and takes necessary measures to implement the best alternative for optimally utilizing the resources of an organization.

EXERCISES

1. Define capacity planning.
2. What are the factors affecting capacity planning?
3. Explain short-term capacity decisions and long-term capacity decision.
4. Explain the process of capacity planning.

❑❑❑

CHAPTER 4

Facility Location

4.1 NATURE OF FACILITY LOCATION

Facility location deals with questions such as where should our main operations be based? Should it be Karnataka, Tamil Nadu or Maharashtra? If Karnataka is most desirable should it be at Bangalore or Mysore? Once such a decision is taken, say Mysore, then the organization has to live for a long time with the prospects and problems regarding the raw material, supplies, labours and other resources and also markets that Mysore would present. Once the manufacturing plant has been erected, it is a long term commitment on the capacity created; capacity in terms of machinery and equipment, in terms of arrangement and development of raw material resources, labour skills, market and distribution channels and the adjustment with the environment. Location of facilities for operations is, thus a long term capacity decision which involves a long term commitment about the geographically static factors that affect a business organization. Some changes and adjustments in capacity, qualitatively or quantitatively at the location are always possible, however in most of the cases, it is not easy to change the location of the operations base once it has been established.

4.2 IMPORTANCE OF FACILITY LOCATION

Facility location and layout are matters of basic importance in business. If a firm opt for a poor location, it will have to face many problems like: non-availability of required raw materials easily, power supply problems, problems from local people, problems of waste disposal and many. Location and layout are interdependent. If the location of a new establishment is satisfactory, it will not be difficult to arrange internal facilities, but it will provide space to satisfy the current and future needs of a concern. We should see, therefore, that the location determines the best housing of equipments and activity. It should, however be the aim of planners to see that the site is chosen to fit the requisite of plant construction and processes. Plant location requires the analysis and weighing of many factors. The problem of location may arise in case of expanding companies which may have to consider the suitable locations for their branch plants and also on the existing site for the purpose of expanding facilities. In order to avoid a bad location it is better to have a systematic method with following steps:

1. Define the location objectives and related constraints.
2. Identify the relevant decision criteria (*a*) Quantitative factor like economic factors (*b*) Qualitative factors (less tangible).
3. Relate the objectives to the criteria using appropriate mathematical models/ economic models such as break-even analysis, economic cost models, linear programming and qualitative factor analysis.
4. Do field research to generate relavant data and use the models to evaluate alternate locations.
5. Select the location that best satisfies the criteria.

4.3 FACTORS OF LOCATION

Decision regarding location requires a careful balancing of several factors. Some of them are more important and are known as *primary factors,* while the less important one are known as *secondary factors.* Sometimes in real problem, secondary factors may show greater influence in selecting the location.

4.3.1 Primary Factors

1. Availability of raw material: In order to minimise the transport cost of raw material to the industrial plant, the nature of raw material is of great importance. A good deal of economy in transport costs can be achieved if the industries, which use weight-losing materials, are located nearer to the source of raw materials. In case industries uses uniquities material, the material transport cost factor do not influence the location much. In addition to the cost factor, other factors related to material to be considered, while taking a decision regarding the locations are: (*i*) Whether the material available indigenously or to be imported?, (*ii*) Any financial linkage to the supply of raw material, so that the material is available below market price, (*iii*) Reliability and continuity of the source of supply, and (*iv*) The security factor of transport. When two locations are equally suitable, then promoter should think of other materials required for processing and availability of market for by-products.

2. Nearness of market for the finished product: Industries, using pure raw materials are generally located nearer to the market for the product produced. By locating the unit nearer to the market, the transportation cost of finished goods will be minimum. In addition to this factor, the chances of finished good getting damaged or spoiled during transport can be reduced. As the industry is nearer to the market, it can catch a big share of the market and can render quick service to the customers.

3. Availability of fuel and power: The problem of fuel and power can also be solved with reference to the nature of raw materials. Industries, which use coal as the source of power for their industries, are located nearer to the coal bed. Because coal loses, substantial weight during processess. Nowadays as the electricity is generally used as power source, coal factor is losing its importance.

4. Transport facilities: A lot of money is spent both in transporting the raw material and the finished goods. Depending on the size of raw material or finished good, a suitable method of transportation like roads, rail, water or air is selected and accordingly the plant location is decided. The point to be considered here is that the cost of transport must be kept at minimum. This is because, the movement of material will simply add to the cost and nothing to the use/economic value of the material or product.

5. Availability of labour: Another important factor influencing the location of industries is availability of suitable and adequate number of labour at a reasonable labour wages. This particularly true in case of plants like, tobacoo companies and tea estates, etc. However, labour will not attract industry by itself as it is mobile. Any labour intensive companies have to select a location nearer to the source of labour. In case industry needs skilled labour, it has to select a location nearer to the source of skilled labours.

6. Availability of water: In case water is used for processing, as in paper and chemical industries, and is also required for drinking, sanitary purposes. Depending on the nature of plant, water should be in adequate quantity and should be of desired quality.

4.3.2 Secondary Factors

1. Soil and climate: The question of soil and climate is really an influencing factor for units processing agricultural products like tea, coffee and rubber, etc. With the development in the field of heating, ventilating and air-conditioning climate of the region is not a problem these days.

2. Industrial atmosphere: The industrial atmosphere cannot be measured in tangible terms, but it has a very important advantage in selecting a location for a plant. Industrial atmosphere may said to exist where people living at a place think instinctively of industry and learn the intricacies of machines without much effort. This is an added advantage for the growth of industries in a particular area.

3. Financial and other aids: The plant should be located in an area where financial institutions are available to get a part of capital as loan and getting working capital and for other financial aids.

4. Availability of facilities like housing, schools, hospitals and recreation clubs: The site should be nearer to infrastructure facilities, so that the labour will have housing facility at a reasonable cost and they will take advantage of educational institutions to educate their children. For health problems, they can use hospitals. They can relax during off periods in recreation clubs. All the above-said facilities will help in motivating the labour to stick to the jobs, *i.e.,* labour turnover will be at low rate.

5. Momentum of an early start: There are number of places where, to begin with, only one or two factories were started. With the passage of time these places gained importance and attract industries. As the number of industrial units increase, certain facilities are developed in that area. Some of the facilities are banking facility; maintenance units and other service related units will come up. With these, availability of skilled labour also increases.

6. Special advantages of the place: Sometime certain facilities are offered by the state, in case the particular region is selected. Facilities like providing subsidies, tax holidays, etc. In this way state tries to develop the underdeveloped areas.

7. Personal factors: Sometimes, promoters, irrespective of economic factors, prefer to start their unit in their place of origin to develop the area and help the fellow community. A good example for this is Ford Company at Detroit.

8. Historical factors: Factors like personal fancies of entrepreneurs or historical incidents may in many cases lead to the development of a place as the centre of an industry. Example is development of Kanpur as textile industry centre.

9. Political stability: The lack of political stability in a state makes for uncertainty in the attitude of State Governments to industries. In locating industrial plan, it must be seen as to whether the State has a record of political and economic stability.

The above-discussed points are mostly economic in character. They, however, ignore the broader national interest and lead to concentration of industries in a few places. This is an undesirable factor as far as national defense is concerned. If there is a balanced development in all the regions, many industrial problems can be reduced. In this connection both Central and State governments are encouraging entrepreneurs to start their units in underdeveloped areas by giving many incentives. The location pattern of an industry can be measured by the formula:

(Number of workers employed in the industry in a certain area)/(Total number of workers employed in the industry all over the country)

This formula gives us an idea about the degree of concentration of industries in one particular area. If we work out the ratio of number of workers in a particular industry and the total number of industrial workers in an area, it will show us how far that particular place depends on that industry.

4.4 RURAL VS. URBAN LOCATION

After considering the economic factors and working out which particular site among the proposed sites, next thing is to think of whether the site should be in rural area or in urban area? No doubt each one has its own advantages and disadvantages. In view of developing rural areas and underdeveloped areas, nowadays State is encouraging the entrepreneurs to start their units in rural and underdeveloped areas by giving certain incentives. Let us now discuss the advantages and disadvantages of rural and urban sites.

4.4.1 Selecting the Site in Urban Areas

Advantages:

1. Urban places will generally have good transport facilities such as rail, road and airways. It will help the entrepreneurs to save time in getting the raw material from source or sending the product to the market point. Also, it provides transport to the employees.
2. Urban places are good market for products manufactured. Hence the producer can avail this opportunity to sell his goods in the local market.

3. A good mix of labour: skilled/semi-skilled/ trained/experienced is available in urban areas. This is an added advantage for an entrepreneur. He can get suitable labour without any difficulty.
4. Power is easily available. Corporations at a reasonable rate supply water to a certain extent.
5. Good medical facilities, good schools and colleges, banks and financial institutions are available in urban areas.
6. Sometimes for small units, which do not use heavy machines, readily built building are available for rent to start the unit without delay before getting a proper site.
7. Training facilities are available to the workers, foremen and other staff.
8. Services of professional people are available with least difficulty.
9. Many related industries, maintenance units and other facilities are available in plenty in urban areas. Sometimes small units working nearer to big industrial units may serve as ancillary units.

Disadvantages:

1. Land cost is very high and the area one can get may be very small and limited. Hence future expansion may become a problem.
2. Cost of building construction is generally high.
3. Expansion in the same area may be a problem.
4. Local taxes and labour charges comparatively will be high.
5. Labour union problems may exist, which will strain the employee-employer relationship.

4.4.2 Selecting a Site in Rural/Sub-urban Areas

Advantages:

1. Plenty of land available at resonable price. For future expansion of the unit it will be very convenient.
2. Unskilled labours are available in plenty at reasonable wages. If the unit is operating on seasonal basis, then seasonal labours are also available in rural area.
3. As the labour union strength is meager, labour problems are less and employee and employer relations will be good.
4. Municipal and other regulations and taxes are seldom burdensome.
5. Government will give subsidies, incentive and tax holidays.

Disadvantages:

1. It is difficult to get skilled labour and managers.
2. Transport facilities may not be that good compared to urban areas.
3. Availability of continuous power may be problem.
4. Good hospitals, schools, financial institutions and banks may not be available.
5. It is difficult to get ancillary service units.

It is better for an entrepreneur to select a site in a place, which is not purely urban or purely rural. That is, it is better to have a location in sub-urban area.

4.5 LOCATION ANALYSIS

Analysis of location alternatives should consider both *objective factors* such as availability of proper transport facilities, availability of suitable labour, tax considerations, material cost and market potentialities and *subjective factors* like civic amenities, Labour union activities, weather conditions, etc. However, the analysis for the choice of appropriate location/site is based on certain methodology commonly adopted by professionals in the field. The below mentioned analysis is generally carried out. (*a*) Cost analysis, (*b*) Qualitative methods, (*c*) Weight method, and (*d*) Ranking method.

An ideal plant location is one, which results in lowest production cost and least distribution and procurement cost per unit. These costs are influenced by a number of factors as discussed above. The various costs which decides the location economy are those of: land, building or rent of the premises, equipment and machinery, labour charges, water, fuel, power, freight (both in incoming and outgoing), raw material cost and taxes, etc. Besides these costs, community attitude, community facilities and housing facilities are also worth considering. The economic aspect of locations are considered or an economic survey or economic analysis is carried out to decide as to which of the possible two or three preliminary selected locations, is the overall best location.

(*a*) **Cost Analysis:** A cost study is made regarding all the alternative sites, to work out what would be the expected production cost per unit of product or what is the rate of return on investment expected and they are compared to select the optimal site. Cost analysis can be done in three ways. They are: (*i*) Comparative cost analysis, (*ii*) Rate of return analysis, (*iii*) Break-even analysis.

(*i*) ***Comparative Cost Analysis*:** In this method all the costs involved in establishment and operating the plant at the site is listed and total cost is worked out to arrive at the total cost involved. Basing on the total cost plant site is selected. The problems given below will demonstrate the method.

Example 4.1: *An entrepreneur wants to select a location for his enterprise and he has conducted economic survey of two locations. The following is the details of his survey. Suggest him which location he has to select?*

Factors	*Location 1*	*Location 2*
Cost of	₹	₹
Land	1,00,000	90,000
Building	1,200,000	1,300,000
Water	5,000	6,000
Power	15,000	17,000
Labour	1,40,000	1,20,000

Incoming freight	1,20,000	1,10,000
Outgoing freight	1,60,000	1,50,000
Fuel	40,000	35,000
Raw material and other supplies	1,40,000	1,30,000
Taxes	4,000	2,000
Total cost	**1,924,000**	**1,960,000**
Community facilities	good	Excellent
Community attitude	All right	Encouraging
Housing facility	Very good	Good
Cost of living	High	Normal
Community size	Small	Medium

(Community shows that area is rural, sub-urban or urban–which gives a rough idea about land cost, labour cost, housing and other factors.)

When we consider the total cost, we find that the location 1 is better. But other factors indicates location 2 is better. As the cost difference is not much and the plot 2 is having good characteristics, we can suggest to the entrepreneur, that he can select plot 2.

Example 4.2: *In a survey to find out the optimal location for a plant, the entrepreneur has collected the following data. Suggests the appropriate location.*

Particulars	*Location A* ₹	*Location B* ₹	*Location C* ₹
Operating costs per year:			
Transportation cost	1,80,000	1,85,000	1,90,000
Labour	200,000	225,000	250,000
Power	50,000	55,000	60,000
Water	28,000	30,000	32,000
Fuel	50,000	60,000	70,000
Rent	50,000	55,000	60,000
Taxes	10,000	12,000	15,000
Insurance	6,000	6,000	6,000
Miscellaneous	5,000	10,000	8,000
Total operating cost	**584,000**	**638,000**	**691,000**
Construction costs			
Land	150,000	100,000	75,000
Building	200,000	180,000	170,000
Other requirements	50,000	40,000	30,000
Total construction cost	**400,000**	**320,000**	**275,000**
Grand total cost	**984,000**	**958,000**	**966,000**

On the basis of total cost involved, site B appears to be most appropriate site for the plant proposed.

Sometimes, cost comparative chart is represented by graphical means. All cost factors are shown to scale on a bar chart, (one bar for each plant). Comparing the heights of the bars, the decision is taken to opt for a particular site. The bar, whose height is smaller, is selected. Figure 4.1 shows one such chart.

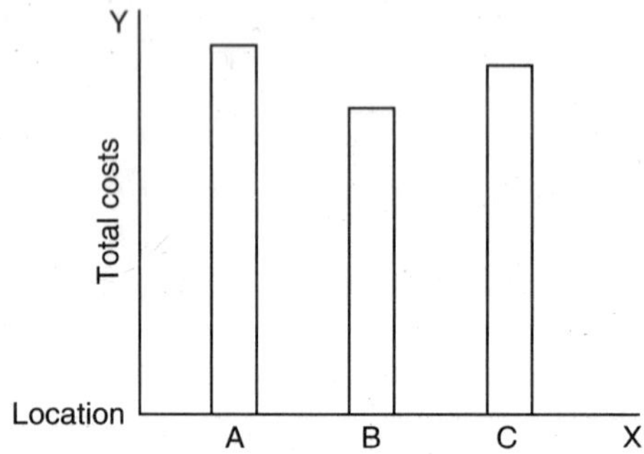

Fig. 4.1 Graphical representation of costs to select a site for a plant.

(*ii*) **Rate of return (ROR) analysis:** This analysis is based on the rate of return on the total investment. Any person who invests money in any business is more interested in knowing what rate of return he is going to get by investing the money? He compares the alternatives and which ever yield higher rate of return, he selects that business.

Example 4.3: *From the following data, select the most advantageous location for setting up a plant to manufacture electric and electronic products.*

Particulars	*Site X*	*Site Y*	*Site Z*
(*i*) Total initial investment	2,00,000	2,00,000	2,00,000
(*ii*) Total expected sales for the period	2,50,000	3,00,000	2,50,000
(*iii*) Distribution expenses	40,000	40,000	75,000
(*iv*) Raw material expenses	70,000	80,000	90,000
(*v*) Power and water supply expenses	40,000	30,000	20,000
(*vi*) Wages and salaries	20,000	25,000	20,000
(*vii*) Other expenses	25,000	40,000	30,000
(*viii*) Community attitude	Indifferent	Wants business	Indifferent
(*ix*) Employee housing facilities	Poor	Excellent	Good

Solution:

Particulars	*Site X* ₹	*Site Y* ₹	*Site Z* ₹
Total expenses: Add (*iii*) to (*vii*)	195,000	215,000	235,000
Rate of Return (R.O.R.) $= \dfrac{\text{Total Sales} - \text{Total Expenses}}{\text{Total Investment}} \times 100$	27.5%	42.5%	7.5%

$$\text{Rate of return of site X} = \frac{2,50,000 - 1,95,000}{2,00,000} = 27.5\%$$

$$\text{Rate of return of site Y} = \frac{3,00,000 - 2,15,000}{2,00,000} = 42.5\%$$

$$\text{Rate of return of site Z} = \frac{2,50,000 - 2,35,000}{2,00,000} = 7.5\%$$

Site Y has higher rate of return and it has good housing facility and community wants business. Hence, it is the ideal site to start industry.

Example 4.4: *An entrepreneur has collected the data about the various costs involved in producing a product and worked out the expected sales, in view to select a plant site. Advise him which is the best site.*

Solution:

Sales and distribution cost	*Location A* ₹	*Location B* ₹	*Location C* ₹
Total initial investment	200,000	200,000	200,000
Expected sales in the year	225,000	300,000	225,000
Distribution Expenses	50,000	40,000	60,000
Raw material and other supplies	80,000	75,000	60,000
Power and Water	20,000	30,000	25,000
Wages and salaries	20,000	20,000	25,000
Miscellaneous expenses	40,000	25,000	30,000
Total expenses	210,000	190,000	200,000
Rate of Return $= \frac{(\text{Total sales} - \text{Total expenses}) \times 100}{\text{Total investment}}$	7.5%	55%	12.5%

Site B is the best compared to A and C.

(*iii*) **Break-even analysis:** Using the break-even analysis concept one may compare objective cost factors. For this one has to identify the fixed and variable expenses for each alternative location and the total cost is worked out. These total cost curves are plotted on a single graph. The location with the lowest cost is selected.

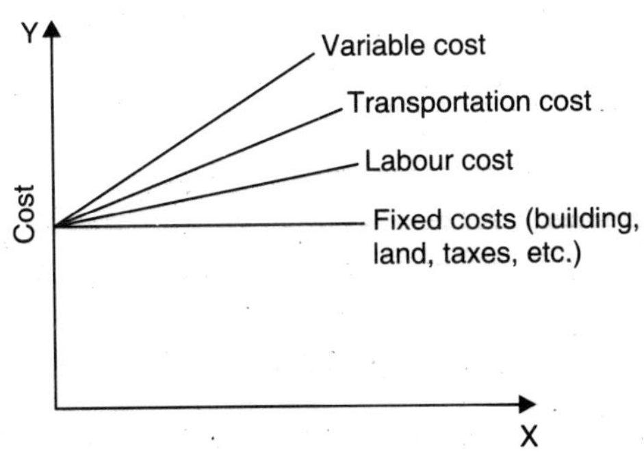

Fig. 4.2 Fixed cost and variable costs of a location.

With expected higher volume compared to total cost as indicated by dashed line in Fig. 4.3, location is selected. If the total cost line for various alternative locations intersect, it indicates that advantage of each location is not absolute but will vary according to the volume of output.

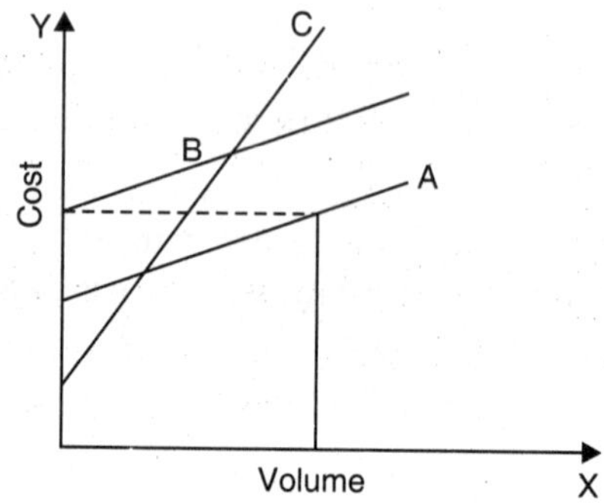

Fig. 4.3 Total cost of alterative sites *A, B, C.*

(*b*) **Qualitative Method:** The entrepreneur, who is making a choice between different site, has to consider intangile factors in addition to the cost factors. Intangible factors, which cannot be measured quantatively, can be expressed as *significant, good, excellent, bad,* etc. Such factors are: housing facility, community attitude, education facilities and medical facilities, etc. The total contribution of intangible factors for each location can be evaluated by the comparative chart as shown below:

Comparative chart for qualitative factors

Factors	*Location A*	*Location B*	*Location C*
Community attitude	Indifferent	Want business	Indifferent
Employee housing facilities	Adequate	Poor	Excellent
Labour supply	Adequate	Adequate	Adequate
Education facilities	Good	Excellent	Insignificant
Recreation	Good	Good	Bad
Union activity	Significant	Significant	Poor
Hospitals	Good	Excellent	Good
Financial institutions	Normal	Available in good number	Normal

Comparatively location B is good.

(*c*) **Weight method:** Here, a weight is allocated to each influencing factor depending on the importance of the factor. Most important factor is given a higher weight and next one is lesser weight and so on. These weight are added and the site, which gets higher total weight, is given preference. One must be careful while selecting a scale for fixing up weight. The following table illustrates the method.

Rating chart for alternative location

Factors	*Maximum weights in units*	*Location A*	*Location B*	*Location C*
Labour cost	250	225	200	150
Nearness to market	200	175	150	125
Nearness to raw materials	400	300	250	200
Water supply	200	100	150	100
Taxes and laws	50	40	45	30
Total weight	**1100**	**840**	**795**	**605**

Location A has got higher weight and this may be recommended. In case two sites are having more or less same weights factor, then other important factors are selected and weights are allocated to select the location.

(*d*) **Ranking method:** In Ranking method, each factor is allocated a rank or a rate according to its importance. Important factor is given a rank of one and all other factors are expressed in whole numbers as multiple of the least number as shown in the table below. Each rank is then multiplied by the appropriate weight factors, to get the score for each location. The score are totalled and the site, which has higher score, is selected.

Relative ranking for different factors

Factors	*Location A*	*Location B*	*Location C*
Labour cost	3	3	4
Nearness to market	2	5	3
Nearness to raw materials	5	1	4
Water supply	4	4	2
Taxes and laws	3	2	1

Total score for each location

Factors	*Weightage*	*Location A*	*Location B*	*Location C*
Labour cost	20	60	60	80
Nearness to market	30	60	150	90
Nearness to raw materials	40	200	40	160
Water supply	10	40	40	20
Taxes are laws	5	15	10	5
Total score	**105**	**375**	**300**	**355**

As the location A has higher score, it is recommended.

EXERCISES

1. Discuss the importance of plant location.
2. Evaluate the advantages and disadvantages of an urban site and rural site.
3. Discuss the techniques of location analysis.

❑❑❑

CHAPTER 5

Facility Layout

5.1 INTRODUCTION

Facility layout is the process of determining the best arrangement of facilities and equipment capable of satisfying anticipated demand with respect to quantity, quality and timing at the lowest cost. It requires meaningful integration of all the elements of conversion process, *e.g.*, men, machines and materials. The facility layout plays an important role in creating an environment conducive to high productivity and the satisfaction of social and psychological needs of people at work. The study of facility layout problem requires specific information about the product or service to be offered, the process to be used and expected level of demand. In addition, it requires the knowledge of constraints imposed by legal, technical or other factor for new facilities. For existing system the proposed layout must satisfy constraints from existing building and other physical structures that form part of production process and the transport network to which it is integrated.

5.2 OBJECTIVES OF FACILITY LAYOUT

- Placing equipment in a position resulting in its maximum utilization.
- Reducing congestion in the flow of materials or workers through successive stages in the process.
- Providing easy access of equipment for maintenance and repair.
- Providing efficient and smooth flow of material within the plant.
- Providing for fast and safe materials handling.
- The best utilization of cubic space.
- The best utilization of human resources by providing a comfortable and safe working environment.

5.3 SYMPTOMS OF A BAD LAYOUT

- Congestion of materials, components and assemblies.
- Excessive amount of work in process.

- Poor utilization of space.
- Long transportation lines.
- Production bottlenecks at certain machines while similar or identical machines have idle times.
- Excessive handling by skilled operators.
- Long production cycles and delays in delivery.
- Mental or physical strain on operators.
- Difficulties in maintaining effective supervision and control.

5.4 FACTORS INFLUENCING FACILITY LAYOUT

The various factors influencing plant layout are as follows:

Material factor: Design, variety, quantity, necessary operations and their sequence.

Man factor: Direct workers, supervision and service help (maintenance) safety and manpower utilization.

Machinery factor: Processing machines, equipment, tools and their utilization.

Movement factor: Inter and intra departmental transport and handling at various operations, storages and inspection, material handling equipment.

Waiting factor: Includes permanent and temorary storages and delays and their locations.

Service factor: Services relating to employee facilities (Such as parking lot, locker rooms, toilets, waiting rooms, canteen, etc.), services relating to quality, production control, scheduling, dispatching, waste control and services relating to maintenance of machinery.

Building factor: Building features, *e.g.,* single and double storey, utility distribution and equipment.

Change factor: Includes versatility, flexibility and expansion.

5.5 PRINCIPLE OF FACILITY LAYOUT

The Six basic principles relating to plant layout as given by Muther are as follows:

Integration principle. It states that all other things being equal, best layout is one which integrates men, materials and machines in the best possible manner.

Flow principle. It states that all other things being equal, best layout is one which makes best use of available cubic space.

Minimum movement principle. It states that all other things being equal, best layout is one which results in minimum movement of men and materials in the plant.

Safety and comfort principle. It states that all other things being equal, best layout is safe and comfortable for employees working in plant.

Flexibility principle. It states that all other things being equal, best layout provides flexibility.

5.6 FACILITY LAYOUT PROCEDURE

The procedural steps involved in plant layout are as follows:

1. Accumulate basic data:
   - Volume and rate of production.
   - Production specifications and bill of materials.
   - Process sheets indicating tools, equipment and methods.
   - Flow process charts.
   - Standard time to complete the operations.
2. Analyse and coordinate basic data:
   - The work force size and type.
   - Number of work stations required.
   - Type of equipment required.
   - Storage and other space requirement.
3. Decide equipment and machinery required:
   - Number of components to be produced.
   - Capacity of each equipment.
   - Time in which a particular work order is completed.
4. Select material handling system based on:
   - Material/products to be moved.
   - Container in which it will be moved.
   - Length of movement.
   - Speed of movement.
5. Sketch the plan of the plot:
   - Mark building outline, roads, storage and service areas, etc.
   - Orientation of plan should utilize maximum natural heat, light and other weather conditions.
6. Determine the general flow pattern:
   - Machinery may be laid as per production process requirements and plant building be erected around.
   - The flow pattern of materials should be such that the distance involved is least between the store and shipping department through the production centres. There should be minimum backtracking and bottlenecks.
7. Design individual work stations:
   - Each work station is laid for achieving optimum performance of operations, material and space utilization and safety and comfort of employees.
8. Assemble the individual work stations layout:
   - The total layout is assembled in accordance with the general flow pattern and the building facilities.

9. Calculate storage space required:
   - Storage space can be calculated by knowing volume of each stored item, numbers of items to be kept in stores and the time for each item to be kept in stores.
10. Make flow diagrams for work stations:
    - After making flow diagrams the work stations are allocated to their respective areas on the plan.
11. Plan and locate service areas:
    - Service areas such as offices, toilets and washrooms, tool rooms, rest and lunch rooms, cafeteria, dispensaries, power generation areas, parking areas, etc. are planned and located appropriately.
12. Make master layout:
    - Master layout is prepared using templets and models.
13. Check final layout:
    - The final layout is checked for safe and economical material handling, adequate production and production control, plant building and its surrounding, service areas, employees safety and comfort etc.
14. Get official approval of layout:
    - The finally checked layout has to be got officially approved from the concerned authorities.

5.7 TYPES OF LAYOUT

There are three basic types of layouts for manufacturing facilities namely, process, product and fixed position.

(*a*) Process layout

A process layout is the arrangement of facilities and equipment in groups according to the function performed. Different jobs follow different paths through the system, depending on their special processing requirements. Examples of such layouts in manufacturing are machine shops for the fabrication of parts, foundries, repair shops, etc. and in service sector are hospitals universities, banks, etc., which are organized so that their resources are grouped according to the different functions performed. As the facilities in process layout are arranged as per their function, it is also referred to as functional layout. It is shown in Fig. 5.1.

Advantages of Process Layout

1. High utilization rates for human resources and equipment, since disruptions due to breakdowns or lack of supplies tend to be localized and easily bypassed.
2. Well-motivated work force due to be relative independence from an unpaced work flow, pride in the application of considerable skill and experience, and considerable freedom for satisfying personal needs at work beyond the earning of wages.

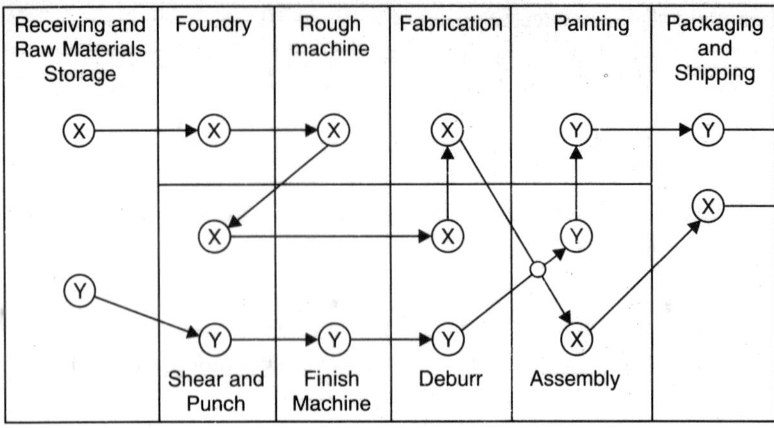

Fig. 5.1 A Process Layout for Producing Products *X* and *Y*.

3. Lower investment in equipment due to the flexibility and interchangeability of general-purpose machines.

Disadvantages of Process Layout

1. High production cost per unit due to the need for special setups and processing requirements with each order.
2. Large in-process inventories.
3. Relatively long processing times.
4. Need for flexible materials-handling equipment (fork-lift trucks, cranes, etc.) to handle numerous product types, shapes and sizes with increased space requirements for travel routes and access to work stations.
5. Complex requirements for scheduling and control since each order must be handled individually

(*b*) Product Layout

A product layout is the arrangement of facilities and equipment in the same sequence as that of the operations needed to complete each unit of the product or service offered. Successive units follow the same path through the system. Examples of product layout applications in the manufacturing are automobiles, TV sets, refrigerators, bottling plants, etc., and in the service sector are cafeteria, insurance claim processing, procedure for issuing passports and driver licenses, etc. As the facilities in product layout are arranged in line as per the processing sequence, it is also referred to as line layout. It is shown in Fig. 5.2.

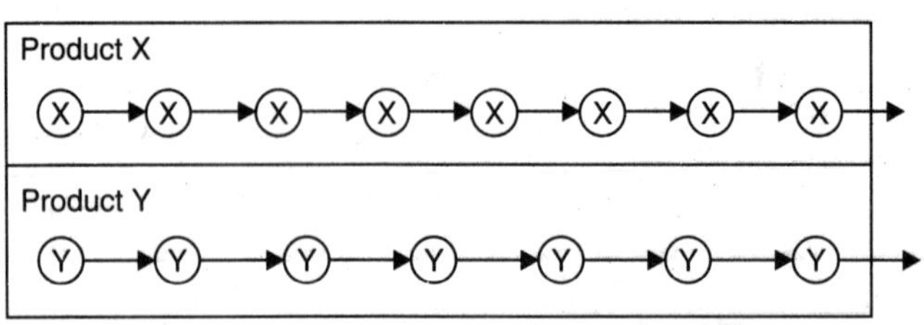

Fig. 5.2 Product Layout for Producing Products, *X* and *Y*.

Advantages of Product Layout

1. Materials-handling time and costs can be reduced through the use of mechanized or automated devices (conveyors, chutes, etc.).
2. In-process inventory requirements are reduced due to the smooth and relatively fast movement of units through the various processing stages.
3. Space utilization is increased due to the elimination or reduction of the need for in-process storage requirements and for flexible materials-handling equipment (fork-lift trucks, carts, etc.) which require special access routes to work stations.
4. Planning, scheduling and control of operations are simplified since these functions have largely been incorporated in the design of the production line. This significantly reduces the need for information processing and decision making.

Disadvantages of Product Layout

1. Lack of flexibility is a key problem when the need arises for adjustments in the line to accomodate desired changes in product characteristics. This often means costly and time-consuming setups or even replacement of expensive equipment.
2. Duplication of identical special equipment for different production lines often requires considerably higher investment than would be necessary if such equipment could be shared by similar products.
3. A breakdown of any one of the several machines on a production line usually forces idleness on the entire line, with costly delays, high maintenance costs, or both.
4. Highly simplified and repetitive tasks and a fast production pace create serious psychological problems with the work force, expressed in the form of low morale and high absenteeism and worker turnover rates.

Table 5.1 provides a comparison of process and product layouts with respect to their nature, application and characteristics.

Table 5.1: Comparison of Process and Product Layouts

	Layout by process	*Layout by product*
Nature	All similar facilities grouped together	Sequence of facilities derived from needs of product
Application	Low quantity production	Large quantity production
	Large range of products	Only a few products
Characteristics	Permits specialist supervision	Little specialized supervision required
	High work in progress	Minimum work in progress
	High material handling cost	Minimum material handling cost
	Ease of provision of services	Difficult to provide services
		Minimum space required
	Can tolerate breakdowns	Single breakdown stops all machines
	Easy to incorporate inspection	Difficult to incorporate inspection
	Individual bonus possible	Group bonus
	Flexibility, variety and product changes possible	Little variety possible
	Possibility for loss or neglect of some job items	

	Maintenance easy Production control complex Production planning simple Long production time	Maintenance out of production hours Production control simple Production planning complex Low production time.

(*c*) Fixed Position Layout

A fixed position layout is the arrangement of facilities and equipment so that workers, equipment, materials, etc. flow to the item being produced or serviced. Example of this type of layout occur in heavy construction (dams, bridges, sky scrappers) air craft, and ship building, etc. It is shown in Fig. 5.3.

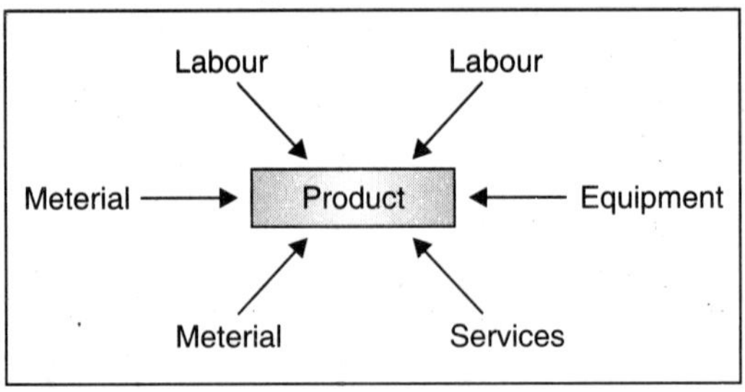

Fig. 5.3 Fixed Position Layout.

This layout is suitable when (*i*) Material forming or treating operation requires only tools or simple portable machines, (*ii*) Only one of the few pieces of an item is to be manufactured. (*iii*) The cost of moving the major piece of material is very high, (*iv*) Skill of the workmanship lies in the abilities of the workers or when it is desired to fix the responsibility of the product quality on the workmen or a group of workers.

Advantages

1. Material movement is reduced to a minimum.
2. The task is usually carried out by a group of operators hence continuity of operations and responsibility is ensured.
3. Production centres often operate independently of each other, and effective scheduling can be planned to ensure minimum total production time.
4. It is more flexible in all aspects.

Disadvantages

1. Movement of machines and equipment to the production centre may be costly and time consuming.
2. Positioning of the material or object or machines may be cumbersome and costly.
3. Machines and equipment utilization is low, owing to handling and positioning time, even if schedule provide for successive employment of the equipment at several production centres.
4. Highly skilled worker are required.

(*d*) Combination of Process and Product Layout or Matrix Layout

A combination of process and product layout to a large extent takes the advantages of both the layout into account. Moreover, these days pure product or process layouts are rare. Most of the manufacturing sections are arranged in process layout with manufacturing lines (layout line) whenever necessary. A combination layout is feasible where an item is being made in different types and sizes. In such cases machinery is arranged in a process layout but process grouping (a group of number of similar machines) is then arranged in a sequence to manufacture various types and sizes of products. The point to note that, no matter the product varies in size and type; the sequence of operations remains same or similar. Figure 5.4 shows one such layout. The extension of this is the *Group layout or group technology,* which is very common in present days.

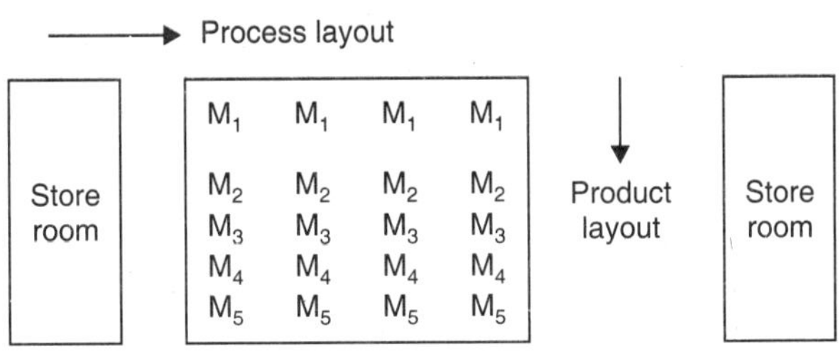

Fig. 5.4 Combination of Process and Product Layout.

A combination layout is also useful when a number of items are produced in same sequence but none of the items are to be produced in bulk and thus no items justifies for an individual and independent production line. (**Example:** files, hacksaws, circular metal saws, woods saws, etc., are manufactured by using this type of layout).

(*e*) Group Layout

Group layout is the type of plant layout in which the plant is divided into groups, each of which contains all the plant necessary to carry out all the operations applied to a given *family* of parts. The parts in the *group family* are generally similar in form, but the essential criterion is that they can all be completed in the special plant group, laid out for that purpose. It will be noted that a change from *functional layout* or *process layout* to *group layout* changes the emphasis from *process specialisation* to *product specialisation.* Each group consists of a set of machines, which can process a wide range of products of different sizes, each requiring processing on the same group of machines, may be in same sequence or different sequence. Figure 4.5 shows a group layout.

Group 1	*Group 2*	*Group 3*
Lathe	Grinding	Milling
Milling	Shaper	Lathe
Shapes	Milling	Grinding

Fig. 5.5 Group Layout.

EXERCISES

1. What do you mean by plant layout? List the objectives of a plant layout.
2. Differentiate product layout and process layout.
3. You want to set up a production plant that produces electronic goods such as refrigerators, micro ovens and air conditioners. What kind of layout will you choose? State reasons.
4. Explain in detail the project layout.
5. Explain group layout.

❑❑❑

CHAPTER 6

Job Design and Work Study

6.1 INTRODUCTION

Job design has a significant impact on both the employees and the organization. Poorly designed jobs often result in boredom, increased labour turnover, reduced motivation, low job satisfaction and increase in organisation costs. These negative consequences can be avoided or minimised through the proper job design. The job design process emphasises the design or redesign of jobs to incorporate factors which lead to the fulfillment of both employee and organisational objectives.

6.2 MEANING AND OBJECTIVES OF JOB DESIGN

Job design is "a process which integrates work content (tasks, functions, relationships), the rewards (extrinsic and intrinsic), and the qualifications required (skills, knowledge, abilities) for each job in a way that meet the needs of employees and the organisation". Some jobs are routine because the tasks are standardized and repetitive; others are non-routine require a large number of varied and diverse skills. Routine jobs constrain employees by requiring them to follow very precise procedures while non-routine jobs allow employees freedom in doing their work. Some jobs are most effectively accomplished by group of employees working as a item whereas other jobs are best done by individuals acting essentially independently. The jobs are differ in the way their tasks are combined, and different combinations create a variety of job designs in the organisation.

The objectives of job design are:

(*i*) To satisfy the needs of the individual employees like interest, challenge, achievement or accomplishment.

(*ii*) To meet the organisational requirements such as higher productivity, operational efficiency, quality of product/service.

(*iii*) To integrate the needs of the individual with the organisational requirements.

6.3 APPROACHES TO JOB DESIGN

Various approaches to job design are:

1. Traditional Approach

Under this approach, a company can allocate duties and responsibilities consistent with the common practices and tradition. This approach to this design of work has many advantages. *First,* it considerably simplifies recruitment, selection and determination compensation. *Secondly*, it is consistent with the employees' expectations and formal training and education programmes as well. *Third*, it is easier to implement since traditional organisation of work has developed over many years and it has been acceptable to many organisations.

2. Scientific Management or Engineering Approach

Under this approach, attempt is made to simplify and standardise the job in order to make it possible for all employees to meet pre-determined level of production. This approach has its origin in the Scientific Management movement which popularised time and motion studies. F.W. Taylor, the father of scientific Management, had an enormous impact on the design of jobs in industrial organisations. Taylor, an engineer, felt that the practice of management should be as analytically based and scientifically grounded as the practice of engineering. For Taylor, the problem was quite simply one of discovering the scientific principles that underlay the practice of management.

Scientific management emphasised the analysis of jobs, and the simplification and standardisation of the tasks performed. As a consultant to the Bethlehem Steel Corporation and other firms, Taylor was able to implement many of his ideas about job design. He and his colleagues developed the process of time and motion study, variations of which are still used by industrial engineers. Through time and motion study followed by work simplification and standardisation, plus the introduction of piecework incentive systems, taylor was able to demonstrate large increases in worker productivity.

The thrust of scientific management was to simplify jobs for the economic benefit of the organisation. Specialisation would result leading to greater productivity if the jobs are simplified through job redesigning. Further, selection and training procedures would also be less elaborate and less expensive.

While appreciating the usefulness of time and motion studies, critics argue that design of work should not be controlled by the simple notions of efficiency, simplification and standardisation. "The psychological implications of time and motion studies are very important. Whenever they are overlooked as they often are, these studies become ineffective tools in the hands of fools". Their contention is that human beings should not be treated like machines; their psychological needs must be taken care of while designing jobs.

3. Human Relations Approach

The human relations approach arose almost as a direct result of the harshness imposed by supervisors who excessively used scientific management principles. An outgrowth of the famous Hawthorn Studies conducted during 1924-33, the human

---

* Tavlor. F.W., *Scientific Management,* Harper and Brothers, New York, 1947.

relations approach de-emphasized the technical components of a job and concerned itself with the impact of employee social and psychological needs on productivity.

Originally, the goals of the Hawthorn investigators were to identify elements of the work environment which fostered productivity. Surprisingly, the investigators discovered that the greatest impact on productivity was that of the social integration patterns of the workers rather than environmental conditions like lighting. These findings led to principles advocating the design of jobs of non-authoritarian leadership styles by supervisors and the fostering of effective work groups.

4. Job Characteristics Approach

This approach focuses on job redesign, work structuring, job enrichment and so on to improve quality of worklife of workers. This job characteristics model of Hackman and Oldham states that three key psychological states of a job-holder determine his motivation, satisfaction and performance on the ob. These states are:

(*a*) Experienced meaningfulness—The degree to which the job-holder experiences work as important and worthwhile.

(*b*) Experienced responsibility—The extent to which the job-holder feels personally responsible and accountable for the result of the work performed.

(*c*) Knowledge of results—Information about how well he is performing the job.

When a worker experiences these states on the job, he feels motivated. He works hard to perform well to the extent these states are important to the worker. Therefore, motivation, satisfaction and performance should be integrated in the job design. These psychological states are generated by the following characteristics or dimensions of the job:

(*i*) *Skill Variety:* The degree to which the job requires the person to do different activities so that he can use a number of different skills and talents.

(*ii*) *Task Identity:* The degree to which the job requires the person to do different activities so that he can use a number of different skills and talents.

(*iii*) *Task Significance:* The degree to which the job has a substantial impact on the work and lives of other both inside and outside the organisation.

(*iv*) *Autonomy:* The degree to which the job provides freedom, independence and discretion to the individual in scheduling the work and in deciding the procedures to be used to do the job.

(*v*) *Feedback:* The degree to which the job provides the individual with clear and direct information about job performance and outcomes.

All the job characteristic dimensions have psychological impact on the workers. The first three dimensions affect whether or not workers view their jobs as meaningful. Autonomy determines the extent of responsibility workers feel. Feedback allows for feeling of satisfaction for a job well done by providing knowledge of results to the job-holder.

The three critical psychological states in turn lead to certain outcomes for both the job and the job-holder. The states provide intrinsic motivation only for those people who have a high need for learning and growth on the job. This desire for personal feeling of accomplishment and growth is called growth need serves as motivator.

The job characteristics theory stresses the intrinsic aspects of jobs and maintains that workers will be satisfied if they view their jobs as meaningful, if they are given adequate responsibility, and if they receive feedback regarding their performance. This goal is accomplished by focusing on the above discussed five specific job characteristics when designing jobs.

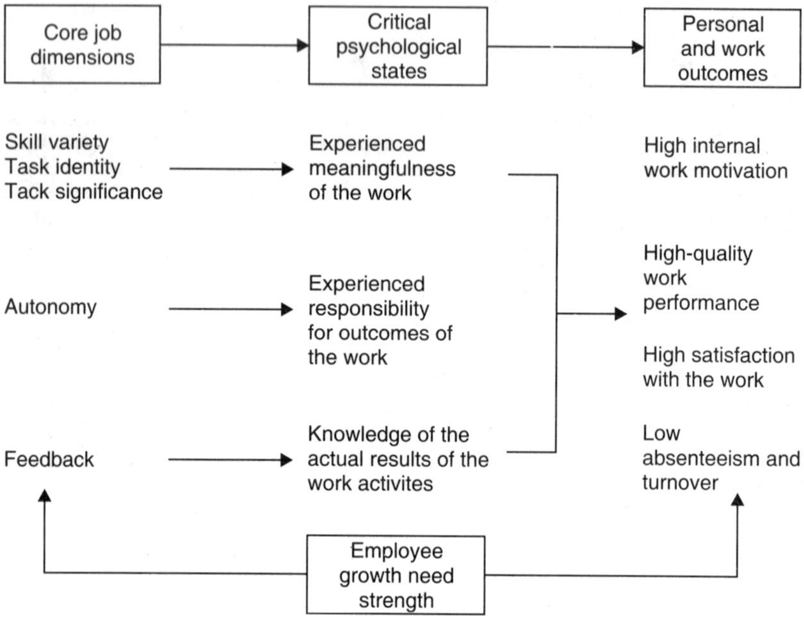

Fig. 6.1 Job characteristics model of Work Motivation.

Ironically, the main feature of the job characteristics design method—its intrinsic psychological motivation—may be its biggest drawback. Supervisors attempting to apply these principles may discover that for many employees these psychological states are unimportant. In fact, research studies indicate that some employees respond exceedingly well to jobs redesigned according to job characteristic dimensions, whereas for others, it has no discernible impact.

5. Socio-technical System Approach

This approach calls for design of work system in such a way that fosters a meshing of the technical and social aspects of jobs. In order to create jobs which have such supportive relationship, work teams not individual jobs, should be studied. Jobs in the traditional sense are non-existent, and instead, combined efforts of employees, supervisors, and union representatives in analyzing significant job operations. Jobs are not necessarily designed to be intrinsically motivating; rather, they are designed so that the work is accomplished. As in scientific management, a supervisor's goal is to ensure that the organisation's objectives are met. However, this is accomplished by concentrating only on critical job aspects, by forming work teams consisting of members who have the necessary qualifications to accomplish the tasks, and by allowing work groups the autonomy to manage their own work-process.

Under the socio-technical system approach, jobs are designed by taking a "holistic" or "systems" view of the entire job situation, including its physical and social environment. The socio-technical approach is situational because few jobs involve identical technical requirements and social surroundings. Specifically, the socio-technical approach requires that the job designer should carefully consider the role of employee in the socio-technical system, the nature of the tasks performed, and the autonomy of the work-group.

The essential elements of the socio-technical system approach are as under:

1. A job needs to be reasonably demanding for the individual in terms other than sheer endurance and yet provide some variety (not necessarily novelty).
2. Employees need to be able to learn on the job and to go on learning.
3. Employees need some minimum area of decision-making that they can call their own.
4. Employees need some minimal degree of social support and recognition at the workplace.
5. Employees need to be able to relate what they do and what they produce to their social life.

6.4 TECHNIQUES OF JOB DESIGN

The techniques of job design and redesign include (1) work simplification, (2) job rotation, (3) job enlargement, and (4) job enrichment as shown in Fig. 2. These techniques are described below:

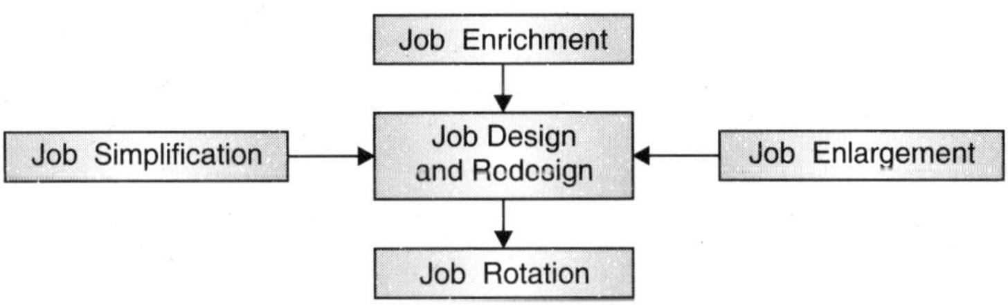

Fig. 6.2 Techniques of Job Design

1. Work Simplification

In work simplification, the complete job is broken down into small subparts, usually consisting of few operations. This is done so that employees can do these jobs without much specialised training. Many small jobs can also be performed simultaneously so that the complete operation can be done more quickly. Time and motion studies are often used for work simplification.

2. Job Rotation

Job rotation refers to shifting people from one job to another within a working group so that there is some variety and relief from the boredom of routine. Hertzberg characterised this approach as merely substituting "one zero for another zero" as it implies horizontal or lateral transfer to a job of the samc lcvcl and status.

3. Job Enlargement

Job enlargement means assignment of additional tasks or duties of the jobs of employees all on the same level. The additional tasks or duties do not require new skills but can be performed with similar skills and efforts as before. In this case, there is enlargement in the horizontal dimension and it may be that the monotonous job remains monotonous only on a larger scale than before. Herzberg characterized this approach as "adding zero to zero" meaning that one set of boring tasks is simply added to another set of boring tasks.

4. Job Enrichment

Job enrichment implies increasing the contents of a job or the deliberate upgrading of responsibility, scope and challenge in work. It is a motivational technique which emphasises the need for challenging and interesting work. It leads to a vertically enhanced job by adding functions from other organisational levels, making it contain more variety and challenge and offer autonomy and pride to the job-holder. Herzberg gave greater emphasis on job enrichment because it is a motivational technique which emphasises the need for challenging and interesting work.

6.5 HUMAN ENGINEERING

Human engineering or Ergonomics is the study of the man-machine system intended to design the machine for its users fitting it to their physiological requirements to minimise fatigue and maximise output. Human engineering means engineering for human use. The human engineer applies scientific knowledge and research methodology to study human areas as they pertain to the operations of the machine systems and concepts. In designing a machine, anatomical, physiological and psychological factors are given due recognition so that the objective to make the work more meaningful and interesting and with less latigue can be fulfilled. This will help the management in securing maximum productivity at the same time giving due regard to human considerations.

6.6 WORK IMPROVEMENT

Work improvement may be defined as scientific techniques of studying and analysing the conditions influencing the quantity and quality of work done by the workers. Some people also call it 'method study'. The term work improvement is very wide and includes the study of work improvement by simplification and standardization of methods, equipment and working conditions. Work improvement is based on the basic philosophy that there is always a scope for improvement in the methods, equipment and working environment. This requires a scientific attitude involving critical analysis of all operations to find the better way of doing the work.

Work improvement is following in industries with the following objectives in view:

(*i*) to eliminate wasteful and inefficient motions.

(*ii*) to remove duplication of efforts.

(*iii*) to improve plant layout and work design.

(*iv*) to improve processes, procedures, machinery, materials and working conditions.

(*v*) to reduce unnecessary fatigue.

6.7 WORK MEASUREMENT

Work measurement deals with assessing the time content of a job performed by an operator to determine the proper time to be allowed and the efforts required for the efficient performance of a job. R.M. Curie has defined work measurement as "application of techniques designed to establish the time for a qualified worker to carry out a specified job at a defined level of performance". Work measurement involves determination of proper time required for the job and that is why it is popularly known as 'time study'. Time study or work measurement is a major component of the work study. Its advantages are as follows:

(*i*) Work measurement determines the normal time for a job and thereby serves as a basis of a sound wage incentive system.

(*ii*) The standard time determined by work measurement helps in labour cost control.

(*iii*) Work measurement provides the relevant data for efficient work planning and control.

(*iv*) Work measurement facilitates effective manning of plant and equipment.

(*v*) Work measurement technique can be useful in reducing the time and cost involved in the proposed production orders.

6.7.1 Work Measurement Vs. Work Improvement

Work measurement is the study and analysis of time taken in the performance of a specific task, work improvement denotes the study of methods and techniques of production for increasing efficiency. These two concepts are interlinked with each other because both of these aim to increasing productivity. Work improvement is done to suggest the best method of doing the job whereas work measurement helps in fixing fair day's work, laying down wage incentive plans, and production planning and control.

6.8 WORK STUDY

Work study is not a new concept. Since the beginning of human race, there had always been a tendency to bring certain improvement in their activities. But this has been recognised at a very later stage. Numerous early pioneers who contributed to the emergence and development of work study are: Frank B. Gilbreth, Fredrick W. Taylor, Beduax, Henry L.Gantt, Marrington Emerson, Halsey, Rowan and others.

Work study is one of the most important management techniques which is employed to improve the activities in the production. The main objective of work study is to arrest the management in the optimum use of the human and material resources. It has three aspects:

1. More effective use of plant and equipments.
2. More effective use of human effort.
3. Evaluation of human work.

It is defined by British Standard Institute as a management service based on those techniques, particulary method study and work measurement, which are used in the combination of human work in all its contexts and which lead to the systematic investigation of all the resources and factors which affect the efficiency and economy of the situation being reviewed in order to effect improvement.

As per the definition work study is the combination of method study and work measurement that means on one hand we have to examine that method of doing the work and on the other hand, we have to find the time required to do the work. Studying the method is to improve the method of production, which results in more effective use of material, manpower and equipment. Work measurement on the other hand assess the human effectiveness and gives guidelines to improve the planning and control methods also serve as the basic of sound incentive scheme. These two techniques are used for optimal utilization of available resources and to improve productivity.

6.9 PRODUCTIVITY AND WORK STUDY

Productivity is the relationship between inputs used and output produced. As far as inputs are concerned, we have to take all the resources that go into production, such as material, man-hour, machine hours, money etc. Therefore, when we talk about productivity, the question arises whether it is material productivity, machine productivity or labour productivity. Hence, while calculating, if we take one particular resource, then we say it the productivity of that particular resource. For example, if we consider the total man-hours used to produce the goods, we say that it is the labour productivity and so on. If we consider all the resources put together and calculate productivity, it is total productivity. Therefore, we can say that the factors affecting productivity are:

1. Nature and quality of raw material.
2. Basic nature of process employed.
3. Efficiency of plant and equipment employed.
4. Volume, quantity, continuity and uniformity of production.
5. Utilisation of manpower.

Productivity can be improved by any of the following:

1. By improving the basic process by research and development or by providing more and improved physical means of production. These needs lot of time to implement and they are considered long-term strategies.

2. It can also be improved by simplifying and improving the product and reducing the variety, which enable us to save scarce resources and improve the use value of the product also reduce the cost of the product. This though not takes that much of time as above but needs considerable time and hence considered medium term strategy.

3. Productivity can also be improved by improving the method of operation and by increasing manpower efficiency at all levels and practising effective planning and organization. This can by achieved by the following scientific management in day-to-day work; hence this is considered a short-term strategy. One of the technique used for improving manpower utilization is work study technique.

6.10 METHOD STUDY

Method study may be defined as "a systematic procedure for analysing the existing method of doing a job including the various human movements involved in it and to evolve the easier, more effective and the most economical method of doing the job." It deals with improving the method of doing work, finding a better way, doing a better job with less effort, and in less time.

The procedure of method study involves the following steps:

1. Identification of the job to be studied.
2. Collection and recording of data about the existing method.
3. Critical analysis of data collected.
4. Development of most practical and economic method.
5. Implementation of new method and evaluating its efficiency.

6.10.1 Techniques of Method Study

There are many techniques which are used for method study but various charts are frequently used for the purpose of method study. Such charts are outline process chart, flow process chart, two hand process chart and multiple activity chart. Figure 6.3 shows symbolic representation of activities.

Symbolic Chart

S.No.	*Activity*	*Symbol*	*Resoulty*
1.	Operation	○	Accomlishment
2.	Transport	→	Movement
3.	Inspection	□	Quality control
4.	Storage	▽	Stocking
5.	Delay	D	Waiting, Holding

Fig. 6.3

1. *Outline Process chart:* This chart is used for recording operations and inspection. This chart is mainly used in the initial state, to know the overall position of the problem and to assess the problem to know whether method study is necessary or not. This chart will only record ○ and □. This chart is shown in Figure 6.4.

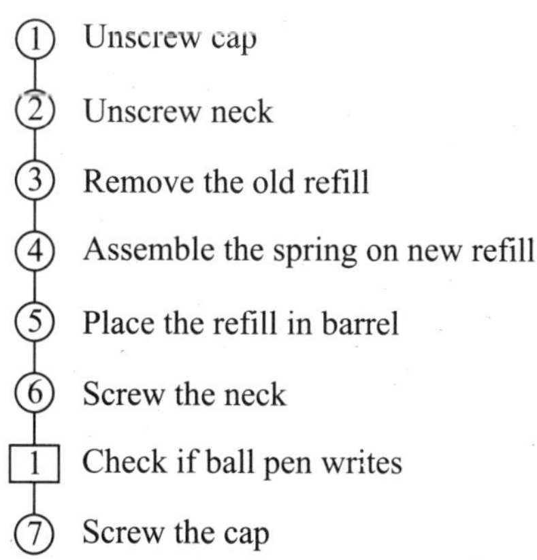

Fig. 6.4

2. *Flow Process chart:* This chart is a graphic representation of the sequence of all operations, transportations, inspections, delays and storages occurring during a process or procedure, and also includes information considered desirable for analysis such as time required and distance moved. Various types of flow charts are:

(*i*) *Material Flow Chart.* This chart is a symbolic representation of all the activities performed on materials. Materials are subjected to all the five activities shown in Figure 6.5.

Activity	*Operations*	*Distance moved (metres)*	*Time (min.)*	*Remarks if if any*
Castings lying in foundry store	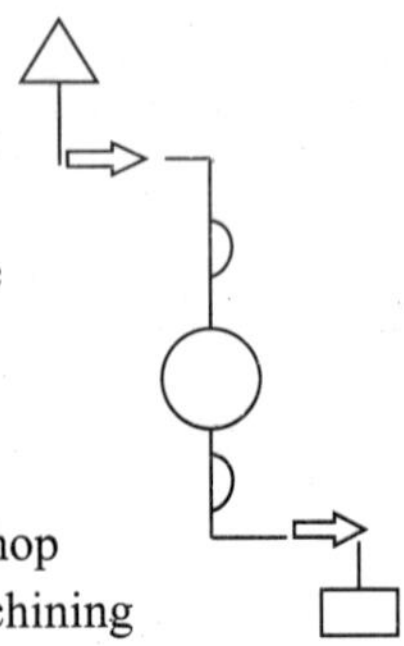	–	–	–
moved to gas cutting machine		10	3	By trolley
Wait cutting machine being set		–	5	–
Rises out		–	20	–
Wait for trolley		–	10	–
Moved to machine shop		20	6	By trolley
Inspected before machining		–	15	–

Fig. 6.5 Flow process chart (man type).

(*ii*) *Man Flow Chart.* Man flow chart is a symbolic representation of the activities performed by a man. There is no representation of storage activities because human efforts cannot be stored. This chart is shown in Figure 6.6.

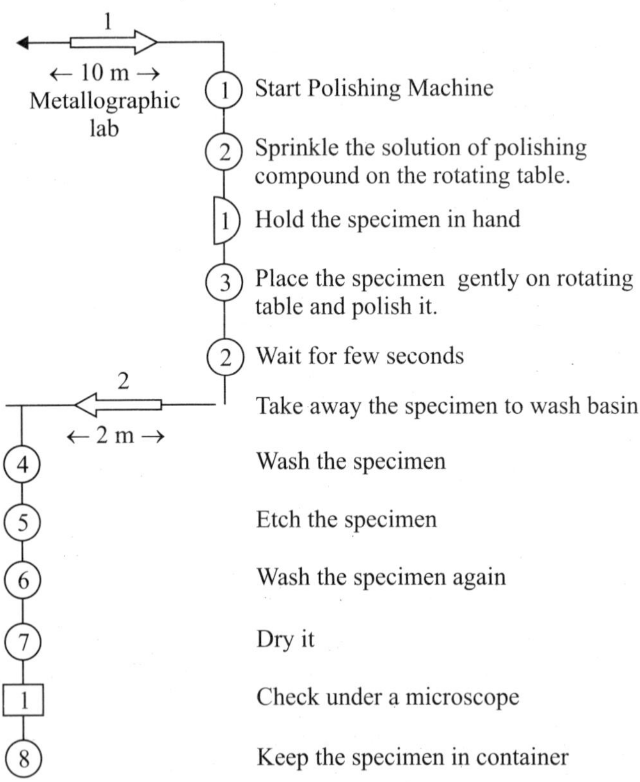

Fig. 6.6

(iii) *Man-Machine Chart.* It is possible to show on a chart the activities of both operator and machine simultaneously. Such a chart is known as man-machine chart.

3. *Two Hand Process Chart:* This chart records activities of both hands of worker as shown in Figure 6.7.

Job: Assembly Nut and Bolt			
Left Hand	*Symbols* L.H.	*Symbols* R.H.	*Right Hand*
Pick up bolt	○	D	Idle
Hold	▽	○	Pick up Nut
Hold	▽	⇨	To left hand
Hold	▽	○	Assemble (screw up)

Summary:

L. H.		R. H.	
○	1	○	3
▽	3	▽	–
⇨	–	⇨	1
D	–	D	1

Fig. 6.7

4. *Multiple Activity Chart:* Multiple activity chart records the activities of men, machine or both on a common time scale in the form of a bar chart. This chart is generally drawn on a graph sheet (Figure 6.8).

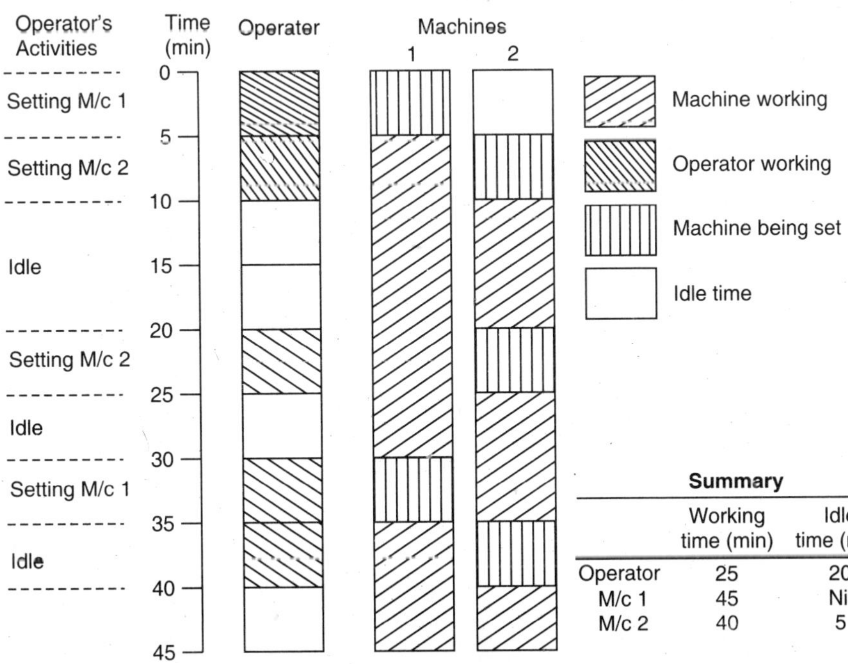

Summary

	Working time (min)	Idle time (min)
Operator	25	20
M/c 1	45	Nil
M/c 2	40	5

Fig. 6.8

(*i*) Outline process chart

Task: Changing Refill of a ball point pen.

(*ii*) Flow process chart (Material type)

(*iii*) Flow process chart (Man type)

Job: Polishing the specimen for metallographic study.

(*iv*) Two Handed Process Chart

(*v*) Multiple Activity Chart

6.11 MOTION STUDY

Motion study is a scientific way of determining the best method of doing a work with the help of a close scrutiny of the motions made by a worker or a machine. Motion study was initiated by Frank Gilberth and his wife Lillian Gilberth. Gilberths defined motion study as science of eliminating wastefulness resulting from unnecessary, illdirected and inefficient motions. Motion study can be divided into three components, namely: (*i*) analysis of therbligs, (*ii*) micromotion study and (*iii*) principles of motion economy.

6.11.1 Therbligs

A therblig is a small part of a job. Therbligs are used to describe the basic elements of movements or fundamental hand motions of the work cycle. Gilbreths gave a list of 17 therbligs. Every therblig is represented by a symbol, a definite colour and with a word. The list of 17 therbligs of motion study is as follows:

Sl. No.	*Therblig*	*Symbol*	*Colur*	*Word*
1.	Assemble		Violet	A
2.	Disassemble		Light violet	DA
3.	Avoidable delay		Lemon yellow	AD
4.	Unavoidable delay	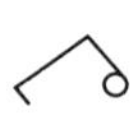	Yellow	UD
5.	Transport Loaded		Green	TL
6.	Transport empty		Olive Green	TE
7.	Search		Black	SH
8.	Plan		Brown	PN
9.	Rest		Orange	R
10.	Pasition		Blue	P
11.	Find		Grey	F
12.	Inspect		Burnt ochre	I

13.	Preposition		Pale Blue	PP
14.	Grasp		Red	G
15.	Use		Purple	U
16.	Hold		Gold ochre	H
17.	Select		Light Grey	ST
18.	Release Load		Carmine red	RL

6.11.2 Micromotion Study

Micromotion study is a study of the fundamental elements of an operation with the help of a high speed movie camera in order to eliminate the unnecessary motions involved in the operation and balancing the necessary motions. The motions are finite sub-divisions of the work cycle. They are expressed in the units of TMU, where one TMU = 0.0006 minute. These motion elements were classified into 17 basic elements known as therbligs.

Micromotion study uses a motion picture camera and a large clock marked off in hundredths of seconds to analyse the motions of workers. The speed of camera should be 1,000 frames per minute. Finding out the number of films that elapse in doing a particular element of motion and knowing the speed of the camera, we can know the exact time value that could be assigned to a particular element. This value can be indicated on the frame scale or time scale drawn on the simo chart (simultaneous motion cycle chart). The study is known as memo motion study if it is done with the help of a slow speed camera.

6.11.3 Principle of Motion Economy

A set of rules were designed by Gilbreth in order to develop better methods. A better method of doing a job is one which consumes minimum of time and energy in performing limb (hand, foot, leg, arm, etc.) motions in order to complete the task and this is possible only, by economising the use of motions.

The rules of human motions as presented by Gilbreth were rearranged and amplified by Barnes, Lowry, Maynard and others. The various rules are as follows:

1. Rules Concerning Human body

(*a*) Both hands should be used for productive work.

(*b*) Both the hands should start and finish their motions at the same time.

(*c*) Except for the rest period, the two hands should not be idle one time.

(*d*) Motions of hands or arms should be symmetrical, simultaneous and opposite.

(*e*) Motions should be simple and involve minimum number of limbs (of course depending upon the type of work). The purpose is to perform the work in the shortest duration of the time and with minimum of fatigue.

(*f*) It is very desirable for a worker to employ momentum to assist himself where the same (momentum) is not to be overcome or conquered by his own muscular efforts.

(*g*) Motions should be smooth and continuous; they should not involve frequent stops and sharp directional changes.

(*h*) Ballistic movements (*i.e.*, as used in driving a nail) wherever feasible should be preferred over controlled because they are easy, fast and more accurate.

(*i*) A worker may use mechanical aids to assist him to overcome muscular effort.

(*j*) Work movements should be rhythmical and automatic, if possible.

2. Rules Concerning Work Place Layout Material Handling

(*a*) There should be a definite, fixed and easily accessible location for materials and tools.

(*b*) As far as possible, materials, tools and other mechanical devices should be kept close to the work place.

(*c*) Gravity should preferably be employed for delivering materials at the work place.

(*d*) An assembled or final product should preferably be dropped on a conveyor (or chute) near the work place, so that gravity and not the operator delivers the job at the required place. Hands should normally not be employed for non-productive work.

(*e*) Tools and materials should preferably be located in the order or sequence in which they will be required for use. It reduces mental strain (on the operator) and the process more or less mechanical.

(*J*) Good illumination is necessary for proper seeing, fast operating and reducing accidents.

(*g*) In order to impart rest to some of his limbs, an operator may sometimes sit or stand while working. This necessitates a relationship between his chair height and the height of the table or work place.

(*h*) In order to reduce fatigue, the seating arrangement of the worker should be comfortable and adjustable.

(*i*) All heavy parts should be lifted by mechanical devices.

3. Rules Concerning Tools and Equipment Design

(*a*) Jigs, fixture and foot operated devices should be employed to reduce the work load on the hands.

(*b*) Wherever possible, those tools should be used which can perform more than one operations. This saves a lot of time, otherwise wasted in searching and picking a number of tools. A tool shown in Fig. 6.7 can carry out three operations, namely (*i*) hammering, (*ii*) drawing out nails, (*iii*) screw driving.

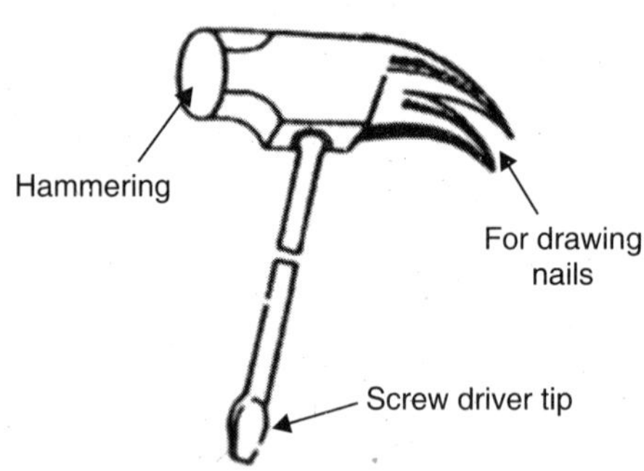

Fig. 6.7 A multipurpose tool.

(*c*) Preferably tools and materials should be preplaced and located near the work place. It saves the time otherwise wasted in searching and bringing the tools for doing job.

(*d*) There should be maximum surface contact between the tools (handle) and the hand. It helps proper application of hand force and minimizes fatigue.

(*e*) Where the work is carried out by fingers (as in typing) the load distributions on-to-each finger should be as per the normal capacity of the finger.

4. Rules Concerning Time Conservation

(*a*) Even temporary ceasing of work, by a men or machine should not be encouraged.

(*b*) Machine should not run idle. It is not desirable that (the lathe chuck holding) the job is rotating but no cut is being taken.

(*c*) Two or more jobs should be worked upon at the same time or two or more operations should be carried out on a job simultaneously. For example, a drill and a turning tool may simultaneously operate on a part being manufactured on a turret lathe.

(*d*) Number of motions involved in completing a job should be minimized.

6.12 DESIGN OF WORK PLACE LAYOUT

Design Considerations

1. Materials and tools should be available at their predetermined places and close to the worker.
2. Tools and materials should preferably be located in the order in which they will be used.
3. Wherever possible gravity should be employed for, the raw material to reach the operator, and finished product to be delivered at its destination. However, it should not be too automatic to become monotonous and bring for the operator. Under such condition suitable breaks and rest periods should be provided.
4. The operator should have comfortable posture. The height of his seat should position the worker such that the work table is about 50 mm below the elbow level to the operator.
5. A worker should have his choice to sit or stand freely during work; that means it should be possible to work both while sitting or standing. A flat foot rest should be provided for the sitting workers. Sitting is less tiring as compared to standing. Seat should not be very long and may have an angle of four degrees. The back of the seat should not restrict the arm movements (essential) during work.
6. Worker should be able to operate levers and handles without changing body position.
7. The work place should have enough illumination, proper conditions of heat, cold and humidity, and reduced dust, noise, etc.

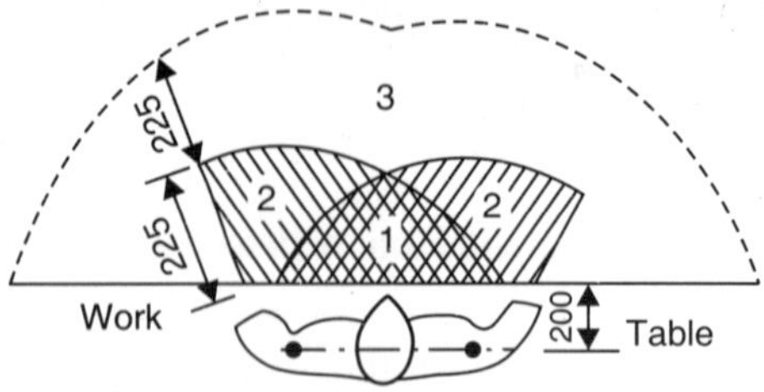

Fig. 6.10 Work place layout showing different areas and typical dimensions (mm).

1. *Actual Working Area:* It is most convenient area for working.
2. *Normal Working Area:* It is within the easy reach of the operator.
3. *Maximum Working Area:* It is accessible with full arm stretch.

Figure 6.11 shows work place layout for assembling small component parts A-l is the actual working area and the place of assembly (POA) where four component, parts P-l, P-2, P-3 and P-4 are assembled together. Bins containing P-l, P-2, P-3 and P-4 and commonly employed tools (CET) (like screw driver, plier, etc.) lie in the normal

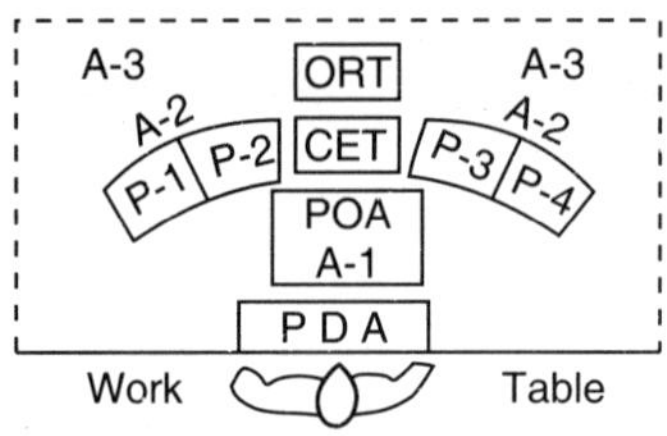

Fig. 9.11 Work place layout for an assembly job.

working area A-2. Occasionally required tools (ORT) (hammers, etc.) lie in the maximum working area A-3. After the assembly has been made at POA, it is dropped into the cut portion in the work table-PDA (place for dropping assemblies)—from where the assembly is delivered at its destination with the help of a conveyor.

This work place arrangement satisfies most of the principles of motion economy.

6.13 TIME STUDY

Time study is an important device of work measurement. It is the "art of recording and analysing systematically the time required to perform a motion or a series of motions." Time study attempts to determine a standard day's work by finding the amount of time needed by workers to perform the various operations. Time study is carried out by studying a worker or a group of workers and carefully measuring with a stop watch how long it takes to perform each operation of the job. It is a must to do a motion study first so that during the time study workers are using only those motions that are necessary and most desirable. If time study is done without motion study and rates are set on the basis of time study, workers may adopt a method of work which will fetch them higher pay. After times have been measured for the various operations of a job, allowance should be made for fatigue and visits to attend to necessary calls. A time study board and a stop watch are used to observe the time required do each operation.

6.13.1 Time Study Procedure

The various steps involved are as follows:

(*i*) Selection of work to be studied.

(*ii*) Establishing standardized methods, equipments and working conditions.

(*iii*) Selecting the average worker who is to be studied while performing the work. Necessary confidence in the worker should be created so as to obtain his cooperation.

(*iv*) Division of work into elements suitable for time study.

(*v*) Studying the operator doing the job. For the validity of results, it is necessary that the analyst should take readings not once but a number of times. The number of times for which time study should be repeated will depend upon the level of confidence needed.

(*vi*) Recording time with the help of stop watch on the time study board for the required number of work cycles.

(*vii*) After the time values for each element for a sufficient number of cycles have been recorded, the mode value is selected. The mode value represents the most frequently appearing time value for an element of the job. Mode values of different elements will be added to get the normal time for doing a job. Normal time is the time required by an average worker working under normal conditions to perform a job.

(*viii*) Adding allowances to normal time to get the standard time. Relaxation allowances include personal allowance, fatigue allowance, delay allowance, etc.

In the above procedure, an average worker should be chosen for the purpose of time study. If an average worker is not chosen, performance stating factor will have to be assessed to determine the normal time for each category of worker. For instance, workers are divided into three categories: (*i*) Poor workers (75 points); (*ii*) Normal or average workers (100 points); and (*iii*) Good workers (125 points). If a good workers is studied, the time taken by him to perform the job will be multiplied by 125/100 to get the normal time taken by an average workers.

Time study board and observation sheet are specially prepared for the purpose of time study. The observation sheet is clamped on the board and a stop watch is fixed in a bracket installed on the right hand top corner of the board. The switching on and off of the stop watch is done by the forefinger of the left hand and the time study data is written by the right hand. The stop watch used for time study is a decimal minute stop watch whose main dial is divided into 100 divisions. The watch also contains a small dial with 30 divisions, each division corresponding to one minute. The time may be recorded either by continuous reading of stop watch or by discontinuous or snap back reading. In the former case, once the stop watch is started, it is not stopped till the completion of the entire job. All the readings are taken continuously. The observer has to subtract the two consecutive readings to get the elapsed time for each element and the button is pressed to bring the taller hand to zero. This method is not very popular because it is susceptible to error in observation.

6.13.2 Advantages and Limitations of Time Study

Advantages:

(*i*) Time study helps in determining the ideal workload of different categories of workers.

(*ii*) The standards of performance evolved as a result of time study may be used for evaluating the performance of employees

(*iii*) Time study helps in designing a suitable incentive wage plan to motivate the workers to increase their productivity.

(*iv*) Cost standards are very accurate if they are based on the results of time study

Limitations:

(*i*) Generally, there are variations of the standard time determined by different observers. Even the same observer sets different standard time each time he is asked to conduct the time study.

(*ii*) Time study involves an element of subjectivity of the observer. Sufficient judgement has to be used by the observer in the choice of a measure of central tendency, deciding the degree of personal allowance and so on.

(*iii*) The standard time determined by time study may not be accurate because of incorrect performance rating of the operator under study.

(*iv*) Time study usually has an adverse effect on the workers. They may not show the normal behaviour pattern when they are being observed. Even the trade unions may resist stop watch time studies.

6.13.3 Time Study Vs. Motion Study

(*i*) Time study is concerned with the determination of time taken by the workers in performing each operation on the job while motion study is concerned with the motion or movements of workers.

(*ii*) Time study covers both workers and machines while motion study covers only workers.

(*iii*) Time study is conducted with the help of a stop watch but motion study is conducted by photographic procedures.

6.14 PRODUCTION AND OPERATIONS STANDARDS

Standards exist at various levels of the organization given below:

1. **Individual Job Standards:** These standards are set for individual. These standards are meant for evaluating individual performance and compensation to workers is given according to their individual performance.
2. **Departmental Standards:** Here several workers may perform as a unit, thus forming a team assembly operations. Standards are meant to evaluate department performance and subsequent supervisor compensation. Standards are set for the department not individual.
3. **Plant Standards:** At the plant, works, or comparable service level unit (such as hospital or a school), a specified volume of goods or service must be produced;

labor, materials and overhead standards must be maintained and at the same time their costs must be controlled.

Uses of Standards

As a basis for making operating decisions, (labour) time standards are used to evaluate the performance of workers and facilities and for predicting, planning, and controlling operations.

Standard cost = Standard usage × Standard labour rate

Actual cost = Actual usage × Standard labour rate

Labour efficiency variance = Standard cost – Actual cost

(*i*) Evaluating individual, department and plant performance

(*ii*) Estimating expenses and revenues.

(*iii*) Evaluating alternatives.

(*iv*) Aggregate planning of work force levels and production rates.

(*v*) Capacity planning and utilization.

(*vi*) Scheduling and sequencing of operations.

(*vii*) Cost estimating of product and production lost.

(*viii*) Planning types of labour skills necessary and budgeting labour expenses.

(*ix*) Formulating standard costs.

EXERCISES

1. Define job design. What are objectives of job design?
2. What are the techniques of job design?
3. Define therblig. Explain principles of motion economy.
4. Define time study. Explain time study procedure.
5. What are the uses of production and operation standards?

❑❑❑

CHAPTER 7

Project Management

7.1 INTRODUCTION

Project management is concerned with defining project activities, maintaining their integrity, and ensuring that they are performed as desired within the allocated time and cost budgets, using organizational and non-organizational resources. Project management is necessary to ensure that a project is completed within the stipulated budget, within the allocated time and performs to satisfaction. The traditional form of organization and management techniques are not suitable for managing project work effectively because of the distinctive characteristics of a project—(*i*) project work is non-routine, non-repititive and constrained by many uncertainties, (*ii*) it involves coordination of the efforts of personnel drawn from different departments and contributions of outside agencies, and (*iii*) the relationship in a project setting are dynamic, temporary and flexible. The various issues relating to project management are project planning, project scheduling, project monitoring and project control. In addition, a satisfactory human relations system is essential for effective project management.

7.2 PROJECT PLANNING

Project planning is a vital aspect of project management. It serves the following important functions:

- Well-defined objectives and policies are formulated to serve as a framework for the decisions to be made throughout the life of the project. The technical and performance objective together with cost standards and time target are specified.
- The activities relating to the project to achieve the objectives are spelt out in detail.
- The activities are properly scheduled and sequenced.
- The resources such as materials, equipment and manpower are allocated to each of the activities on the basis of information available subject to revision at a later date when more information is available.
- The time and cost required to accomplish each activity and to achieve the objective of the project are estimated.

- An organization that can effectively implement, monitor and control the project plan is developed.

7.2.1 Techniques of Project Planning

The project planning techniques include; Work Breakdown Structure (WBS) and Project Execution Plan (PEP).

1. Work Breakdown Structure (WBS)

The work breakdown structure represents a systematic and logical breakdown of the project into its component parts. It is constructed by dividing the project into its major parts, each of these being further subdivided into sub-parts. The process is continued till a breakdown is done in terms of manageable units of work for which responsibility can be defined. The work breakdown structure helps in:

(*i*) effective planning by dividing the work into manageable elements which can be planned, budgeted and controlled,

(*ii*) assignment of responsibility for work elements to project personnel and outside agencies (contractors), and

(*iii*) development of project control and information system.

The work breakdown structure can be based on a function oriented approach as shown in Fig. 7.1(*a*), work distribution as shown in Fig. 7.1(*b*), and hardware levels as shown in Fig. 7.1(*c*).

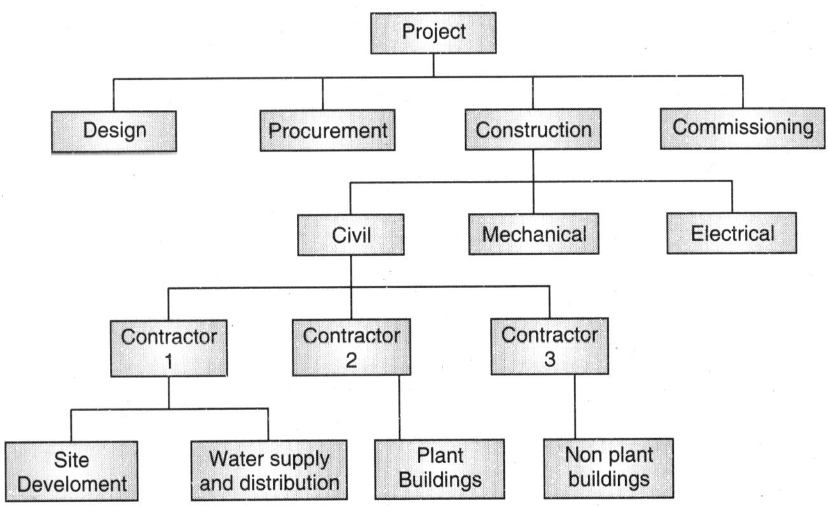

Fig. 7.1 (*a*) Work breakdown structure (function oriented).

2. Project Execution Planning (PEP)

Project execution planning refers to strategic planning to facilitate the development of detailed project planning based on Gantt charts or network planning. It consists of four types of planning as below:

(*a*) ***Contract planning:*** A contract is an aggrement between outside agency and project owner for carrying out specific Jobs related to work.

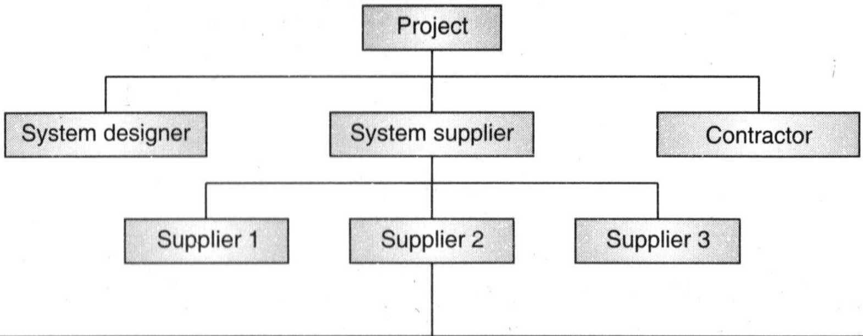

	Design	Supply	Erection	Commissioning
System 1		√	√	
Subsystem 1A	√	√	√	
Subsystem 1B		√	√	
System 2	√	√	√	
Subsystem 2A		√	√	√

Fig. 7.1 (*b*) Work breakdown structure (based on work-distribution).

It involves defining scope of work and responsibility, spelling out terms of payment and identification and protection from risks by the contractor and project owner. Apart from defining terms and conditions of individual contracts, contract planning decides on the number of contracts to be made based on factors such as job speciality, availability of contractors, value of contracts, time and budget constraints, quality of work and work schedule requirements. The National Association of Consulting Engineers (NACE) in India has developed standard forms related to general conditions of contracts to facilitate contract preparation.

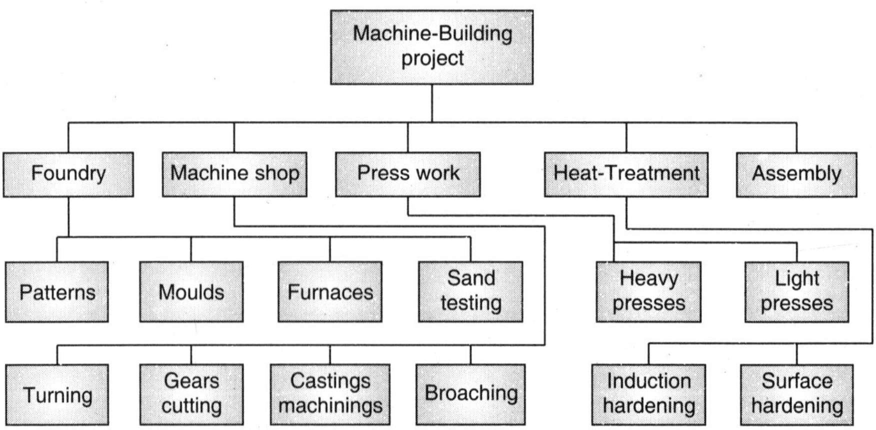

Fig. 7.1 (*c*)

(*b*) ***Work Packaging:*** A work package in a project is a smallest division of work where it retains the characteristics of a project. Work package planning refers to the identification of all the work packages in a project, grouping them or keeping them as they are to form viable work units for entrusting to a single responsibility centre. The benefits of work package planning are as given below:

- Work packaging enables better organization and management of projects. By entrusting of individual work packages to a single responsibility centre, close adherence to time, cost and technical performance targets is ensured.
- Work packaging ensures that all agencies in a project think and channelize their effort towards the completion of work packages in a coordinated manner for the completion of the project.
- Work packaging facilitates the scheduling, budgeting and awarding of contracts on the most competitive terms and in exercising control over time, cost and technical performance.

(*c*) ***Organization Planning:*** Organization planning involves integrating the activities and functions of various departments and outside agencies (contractors) involved in project work under one individual referred to as the Project Manager or Project Coordinator. Depending on the authority given to the Project Manager/ Coordinator, different forms of organization planning are possible. The different forms of project organization are discussed below:

(*i*) *Project Manager Serving as Staff:* In this form of organization, the project manager/coordinator works in a staff position with the primary responsibility of coordinating the work of various people in the functional departments associated with the project work. The project coordinator does not have any authority. He serves as a focal point for receiving project-related information and communicates it to chief executive. The chief executive expects him to promote the cause of the project by rendering advice, sharing information and providing assistance. However, the project manager may find it difficult to exert leadership because he does not have any formal organizational authority. The influence would depend on his professional competence closeness to the top management and persuasive abilities. This is a weak form of project organization and is suitable only for small projects. It is shown in Fig. 7.2.

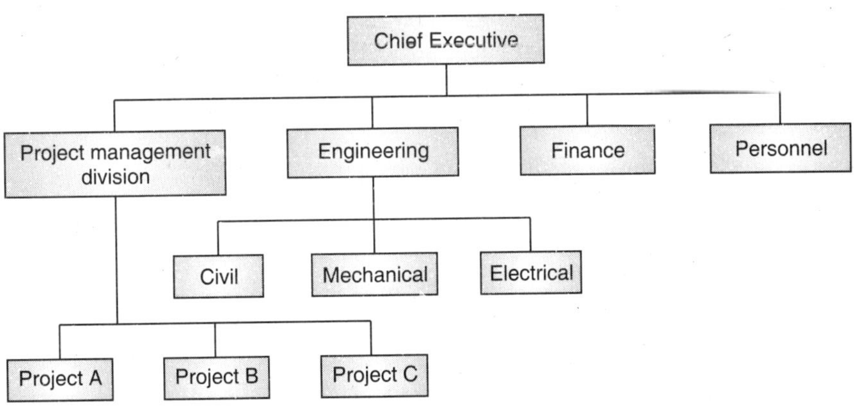

Fig. 7.2 Matrix Organization.

(*ii*) *Project Management as a Separate Division:* This form of organization *provides* for a separate project management division to implement projects. The project management division is interested to coordinate the project work related to different functional departments through project managers in-charge of different projects.

(*iii*) *Matrix Organization:* In a matrix organization, the personnel working on the project have responsibility to their functional superior as well as to the project manager. The authority influence of the project manager cuts across the traditional vertical line of command. The typical matrix organization chart is shown in Fig. 7.3. The project manager integrates the contributions of personnel in various functional departments towards the realization of project objectives. The personnel maintain their departmental affiliation and are responsible to their functional managers, but are also responsible to the project manager. The matrix form of organization is in contrast with the traditional organizational theory; there is a dual subordination, responsibility and authority are not commensurate, the hierarchial principle is ignored. The matrix form of organization involves greater organizational complexity and creates inherently conflicting situation. However, it provides for simultaneous pursuit of twin objectives, (*i*) efficient utilization of manpower resources working in the different departments and (*ii*) effective attainment of project objectives.

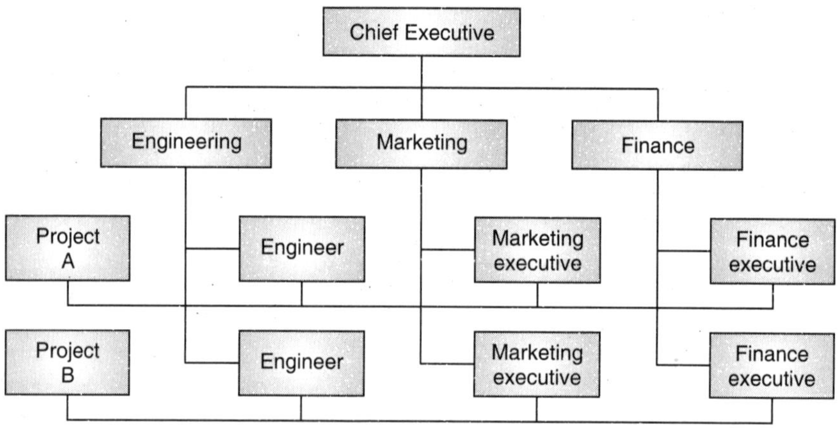

Fig. 7.3 Matrix Organization.

Advantages of Matrix Organization

- Allows flexibility in coordination.
- Broadens participation in sense of team work.
- Enhances job enrichment.
- Provides increased feedback.
- Increases awareness of interdependencies.
- Extends traditional hierarchical relationships.
- Allow individuals to have multiple roles.
- Encourages flexible membership and greater acceptance of change.
- It is useful in complex and large scale projects.
- It is useful in rapidly changing technical environments.

Disadvantages of Matrix Organization

- It may undermine traditional authority.
- It creates dual or overlapping authority.
- It modifies reporting process.

- The multiple leadership may confuse group members.
- It may drain key personnel from other operations.
- The participation involves added duties.
- It increases work responsibilities.
- It is not useful at lower levels or for routine projects.
- It tends to grow and become over extended.
- It highlights problems of communication between technical specialists.
- It stifles creativity if not changed periodically.

(iv) Task Force Organization: In this arrangement the project manager is delegated the full authority to make decisions for the project, but he is required to operate within the functional organization's policies and procedures. There is clearly no intervention from the functional departments, no dual decision making and no dual reporting for the workforce. A task force is created by drawing personnel from the various functional departments and putting them under the project manager. The personnel so assigned continue to receive administrative support from their parent departments but respond only to the project manager. The persons on the task force receive all directions from the project manager, but are required to follow the policies and procedures of their parent departments. In case, the directions of the project manager are contrary to functional policies and procedures, the task force notifies both the functional head and the project manager. The project manager and the functional manager sort out the matter either mutually or through referring it to the chief executive. The loyalty of the functional staff lies with the project work. The functional heads are delegated to staff positions and strong corporate support would be required to ensure adherence to policies and procedures laid by functional heads. The time and cost objective of the project gets best attention. The project moves very fast. It is most suitable when technology is simple and the project is small. It is shown in Fig. 7.4.

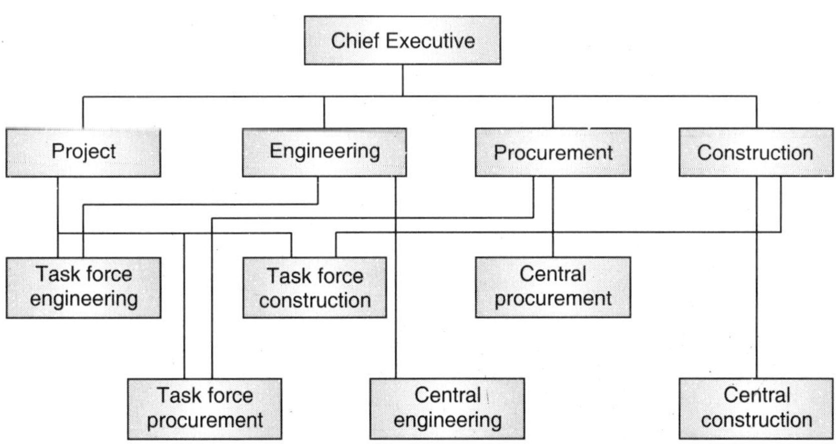

Fig. 7.4 Task-Force project organization.

It is always not possible to create task forces because of shortage of specialists, who can be assigned to task forces. Large an complex projects have complex problems, solutions of which require involvement of multiple specialists of the same discipline. This is not possible in a task force. In large and complex project, the specialists on a task force are senior personnel, who may not be very particular in following their departmental policies and procedures while carrying out the project work. If the duration of the project is long, the specialists deputed on project task force, tend to forget their parent department and become used to project culture leading to a totally projectized organization.

(*v*) *Totally Projectized Organization:* It is autnomous project organization where project manager is chief executive and has total authority.

This arrangement is suitable for very large and complex projects with typical geographical locations. The project manager has to carry out a lot of administrative work in this type of organization. Apart from administering the project work, the project manager also administers functions such as personnel and accounts which are not directly related to the project work. The project manager is interested with administering diverse functions such as engineering, commercial, personnel and accounts. The big companies like Engineers India Ltd. executing huge projects outside the country may opt for this type of project organization. It is shown in Fig. 7.5.

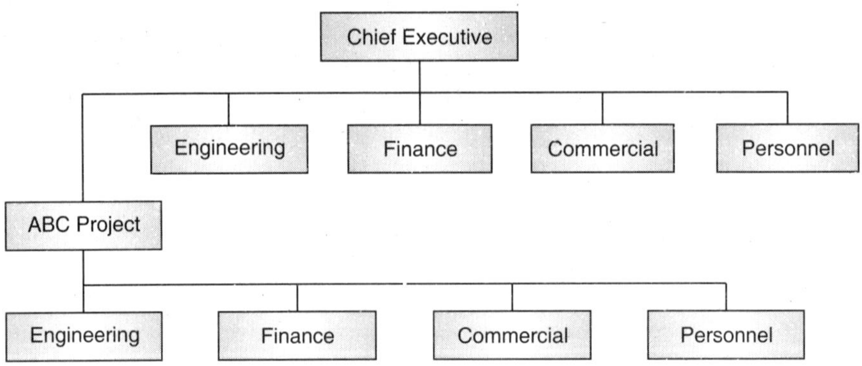

Fig. 7.5 Totally projectized organization.

(*d*) ***Systems and Procedure Planning:*** Systems and procedure planning play an important role for carrying out project work smoothly. Important aspects of system and procedure planning given below need to be formulated before starting of project to avoid confusion during the project.

Contracts Management: It relates to developing procedures related to general conditions of contract, tendering procedure, tender evaluation and award of contract, signing of agreements, measurement of completed work, billing and payment, work inspection and acceptance and takeover of completed work.

Configuration Management: It refers to the efforts made to maintain the integrity of the project in terms of aims, objectives and physical configuration, as finalized before the start of the project. Any change contemplated in the physical configuration of the project at any time during the execution of the project should not alter or adversely affect the aims and objectives of the project. The procedures that require to be formulated relate to the finalization of the project scope in terms of work breakdown structure, finalization of engineering designs, design revision, value engineering and design changes.

Resource Management: Procedures related to men and materials resources need to be carefully formulated for effective utilization of these resources. Procedures related to manpower resources relate to forecasting, recruitment, training and development, performance evaluation, delegation of authority, etc. The procedures for material resources deal with conditions of purchase, ordering procedure, vendor evaluation, insurance and claim procedure, billing and payment procedure, materials status reporting, expediting, etc.

Cost Management: In order to ensure completion of project work within the stipulated cost, the procedures relating to cost management such as project cost estimation, cost coding, cost updating, cost reduction, price evaluation of bids, cost reporting, cost review and cost audits need to be formulated.

Time Management: To ensure project completion within an allocated time, proper procedures related to time management such as preparation of WBS, development of project schedules, progress measurement, project review, revision of project schedules, feedback and updating of schedule and reporting need to be formulated.

Communication Management: Information management is most important aspect relating to project management. Procedures related to communication management deals with correspondence, filing, reporting, recording minutes of meetings, data logging, general information, making presentations, project workshops, etc. which need to be formulated.

7.3 PROJECT SCHEDULING

Project scheduling establishes a time-frame for the various activities of the project. The objectives of project scheduling relate to the shortest time, for completing the entire project, identifying activities which are bottlenecks and must be monitored closely, flexibility or slack available with bottleneck activities, the effects of shifting resources from ordinary activities to bottleneck activities and determining the probability of meeting specific schedule deadlines if the duration of the activities is uncertain.

Project Scheduling Methods

Gantt charts and network scheduling are the two simple and widely used techniques.

7.3.1 Gantt Chart

It shows the relationship of project activities over time. Project activities are plotted against the horizontal time axis. Gantt chart symbols, as given below, are used to denote start of an activity, end of an activity, actual progress of the activity and to depict the point in time where the project is at a given time.

As shown in Fig. 7.6 manufacturing planning is scheduled to start at the beginning of week 1 and end after one week, material procurement is to start at the beginning of week 2 and end after 1 week. The schedule start of an activity is an indicated by an open bracket and a closed bracket indicates the schedule completion. The heavy line indicates the completed portion of the activity. The present time is indicated by an arrow at the top of the chart. This chart can be used to monitor the progress and to adjust the activities accordingly. Figure shows that the progress of materials procurement activity needs to be expedited.

GANTT CHART SYMBOLS

Symbol	*Symbol Meaning*
[	Start of an Activity
]	End of an Activity
[——]	Actual Progress of the Activity
V	Point in time where the project is now

Fig. 7.6 depicts a Gantt chart for scheduling a manufacturing project. The project activities are manufacturing planning, materials procurement, fabrication, assembly and inspection.

Project activities	Week 1	Week 2	Week 3	Week 4
Manufacturing Planning	[——]			
Material Procurement		[——]		
Fabrication			[——]	
Assembly				[ ]
Inspection				[ ]

Extent of Completion__________

Fig. 7.6 Gantt chart for scheduling a manufacturing project.

Advantages of Gantt Chart

- It is simple to understand.
- It can be used to show progress of project activities.

Limitations of Gantt Chart

- It cannot show interrelationship among activity on large complex projects.
- There is a physical limit on the size of the Gantt chart,
- It cannot cope easily with frequent changes and updating.

7.3.2 Network Analysis

Network Analysis is a technique by which projects are broken down to small jobs or events and arranged in a logical sequence to form a network. The network helps in identifying those jobs or events which affect the completion of the project in time. The network-approach to project management is a major advance in management science. PERT/CPM are the two management tools for planning, scheduling and controlling of numerous inter-related activities in a big project.

Origin of PERT and CPM

CPM (Critical Path Method) was developed by E.I. Du Pont de Nemours company in 1956-58 as an application to construction projects and was later extended to a more advanced status by Mauchly Associates. Around the same time PERT (Programme

Evaluation and Review Technique) was developed by the U.S. Navy for scheduling the research and development activities for the Polaris Missile Programme.

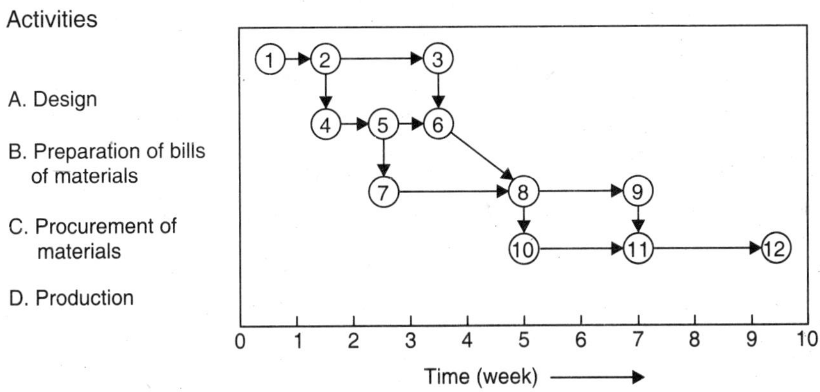

Fig. 7.7 Network.

CPM vs. PERT

There is not much'difference in between CPM and PERT. In fact, both were developed simultaneously and independently for the similar requirements. The only difference lies in the 'event' or 'activity' on which network is constructed. The 'activity' stands for the time-consuming part of the project. It represents a job; on the other hand, an 'event', also called 'node' is either a beginning or an end of a job. The CPM analysis is activity-oriented and PERT is event-oriented. Also there is no allowance for the variation or uncertainties in the estimation of time CPM. However, a common approach can be applied for the analysis of both types of network.

Terms Related with PERT and CPM

(*a*) *Activity:* It is a physically identifiable part of a project which consumes time and resources. In the network, activity is represented by an arrow, the tail of which represents the start and the head, the finish of the acticity.

(*b*) *Event:* The beginning and end points of an activity are called events. It is a point in the time and does not consume any resources. It is generally represented by a number circle. The head even, called the *j*th event has always a number higher than the tail event, called the *i*th event, *i.e.*, $j > i$.

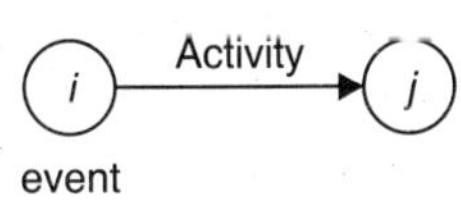

(*c*) *Path:* An unbroken chain of activity arrows connecting the initial event to some other event is called a path.

(*d*) *Network:* It is the graphical representation of logically and sequentally connected arrows and nodes representing activities and events of a project. Networks are also called arrow diagrams.

(*e*) *Rules for network construction:*

(*i*) The length of the arrow does not indicate the time which the activity takes or the resources which the activity consumes.

(*ii*) The direction of the arrow indicates the direction of work flow.

(*iii*) No activity can be shown twice in the network unless it occurs twice in the project.

(*iv*) No two events will be directly connected by more than one activity.

(*v*) An event cannot occur until all activities leading to it are completed.

(*vi*) No activity can begin until its immediate preceding event has occurred.

(*f*) *Dummy:* An activity which only determines the dependency of one activity over the other, but does not consume any time is called a dummy activity. Dummies are usually represented by dotted line arrows. For example, in Fig. 7.8 the start of activity C depends upon completion of activity A and B and the start of activity E depends upon the completion of activity B.

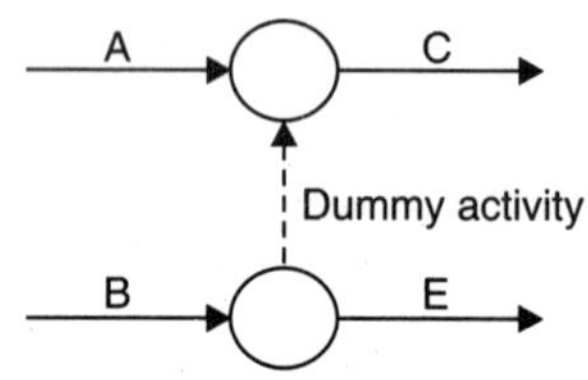

Fig. 7.8

Example 7.1: *The precedence relationship among activities are:*

B > A; C > B; D, E, H > C; F > D and G > E. Draw the network.

Solutions: The project can be represented in the form of a network as shown in Fig. 7.9.

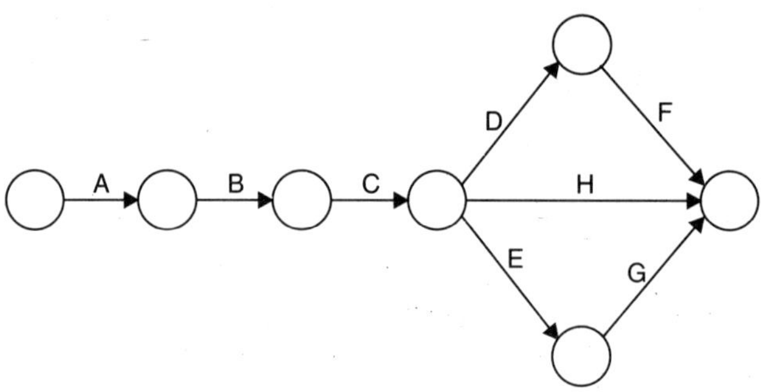

Fig. 7.9

Numbering the Events (Fulkerson's Rule)

After the network is drawn in a logical sequence, every event is assigned a number which is placed inside the node circle. The number sequence should be such as to reflect the flow of the network. The rule devised by D. R. Fulkerson is used for the purpose of numbering. It involves the following steps :

1. The initial event which has all outgoing arrows with no incoming arrow is numbered '1'.
2. Delete all the arrows coming out from node '1'. This will convert some more nodes (as least one) into initial events. Number these events 2, 3,...
3. Delete all the arrows going out from these numbered events to create more initial events. Assign the next numbers to these events.
4. Continue until the final or terminal node, which has all arrows coming in with no arrow going out, is numbered.

To illustrate the numbering technique let us consider the network shown in Fig. 7.10.

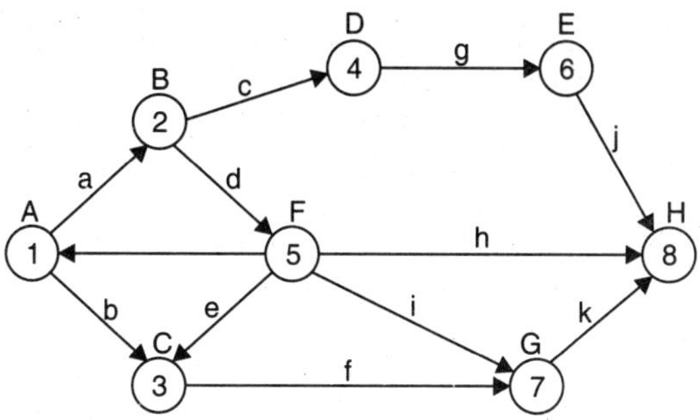

Fig. 7.10

Event A is initial event and is numbered 1. Delete the arrows *a* and *b*. This will create two more (B and C) initial events. Number these 2 and 3. Now delete the arrows coming out from nodes 2 and 3, *i.e.*, delete the arrows *c*, *d*, *e* and *f*. This converts D and F into initial events. Number these nodes as 4 and 5. Then delete the arrows *g*, *h* and *i*, the two nodes E and G become the initial nodes and they are numbered 6 and 7, then the last event or terminal event is numbered as 8.

Time Estimates

The CPM system of networks omits the probabilistic considerations and is based on a *Single Time Estimate* of the average time required to execute the activity.

The PERT system is based on *Three time Estimates* of the performance time of an activity. They are:

(*i*) *The Optimistic Time Estimate:* The shortest possible time required for the completion of an activity, if all goes extremely well.

(*ii*) *The Pessimistic Time Estimate:* The maximum possible time the activity will take if everything goes bad. However, major catastrophes such as earthquakes, floods, storms and labour troubles are not taken into account while estimating this time.

(*iii*) *The Most Likely Time Estimate:* The most likely time is the time an activity will take it executed under normal conditions.

Frequency Distribution Curve for PERT

We have three time estimates for a PERT activity, the optimistic (t_o); pessimistic (t_p) and the most likely time (t_m). In the range from optimistic to pessimistic, there can be a number of time estimates for the activity. If a frequency distribution curve for the activity times is plotted, it will look like the one shown in Fig. 7.11. It is assumed to be a β-distribution curve with a unimodal point occurring at t_m and its end points occurring at t_o and t_p. The most likely time need not be the midpoint of t_o and t_p and hence the frequency distribution curve may be skewed to the left, skewed to the right or symmetric.

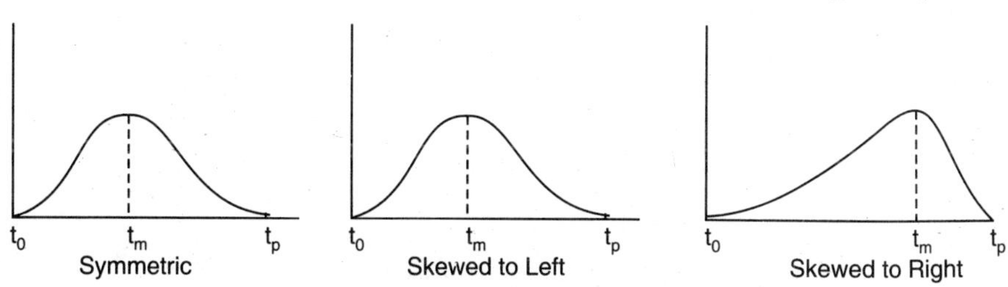

Fig. 7.11 Distribution Curves.

Though the β-distribution curve is not fully described by the mean (μ) and the standard deviation (σ), yet in PERT the following relations are approximated for μ. and σ:

$$\text{Mean, } \mu = \frac{t_o + 4t_m + t_p}{6}$$

standard deviations,

$$\sigma = \frac{t_p - t_o}{6}$$

$$\text{By definition, variance V} = \sigma^2 = \left(\frac{t_p - t_o}{6}\right)^2$$

Expected time or *average time* of an activity is taken equal to mean. This is the time which the activity is expected to consume while executed.

$$t_E = \mu = \frac{t_o + 4t_m + t_p}{6}$$

Example 7.2: *In network shown in Figure 7.12. For each activity, the three estimates t_o, t_m and t_p are given along the arrows in the $t_o - t_m - t_p$ order. Find t_E and variance.*

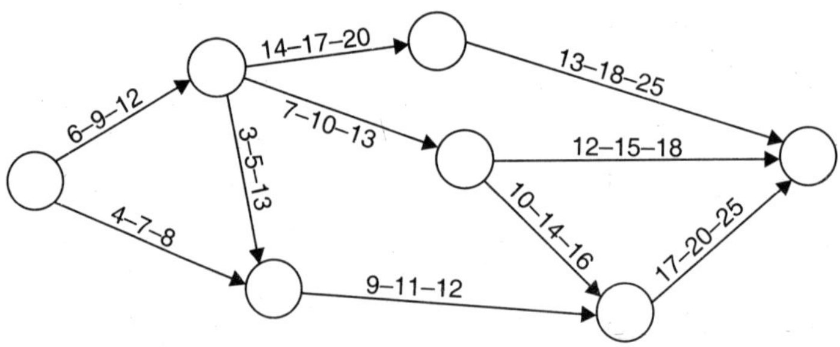

Fig. 7.12

Solution: Let us first number the events. Using Fulkerson's rule, we start from the left and number the initial event as 10. Then, assuming the arrows emerging from node 10 to be deleted, we create one more initial event and number it 20. This procedure is repeated until all the events are numbered.

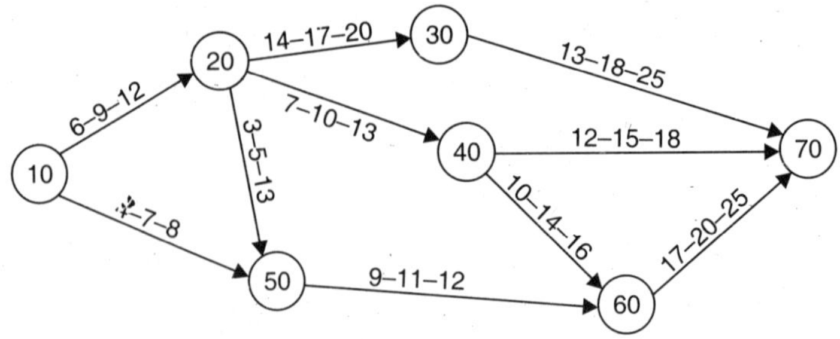

Fig. 7.13

Now we put the events in a tabular form and calculate the variance and expected times. These calculaions can be carried on the network itself also.

Table 7.1

Activity i j		t_o	t_m	t_p	$\sigma^2 = \left(\frac{t_p - t_o}{6}\right)^2$	$t_E = \frac{(t_o + 4t_m + t_p)}{6}$
Predecessor event i	*Successor event j*					
10	20	6	9	12	1.00	9.0
10	50	4	7	8	0.44	6.7
20	30	14	17	20	1.00	17.0
20	40	7	10	13	1.00	10.0
20	50	3	5	9	1.00	5.33
30	70	13	18	25	4.00	18.33
40	60	10	14	16	1.00	13.67
40	70	12	15	18	1.00	15.00
50	60	9	11	12	0.25	10.83
60	70	17	20	25	1.78	20.33

PERT Computation

PERT finds applications in planning and control of complex set of activities, functions and relationships. By controlling the critical activities the project can be completed within the scheduled time limit. PERT is mainly concerned with event.

(*a*) Forward Pass Computations: Calculation begin at the initial event and move towards the end event. Initial event is assigned zero time, and then proceeding to the next event in sequence, the time at which that event is expected to occur at the earliest is calculated. This is called the Earliest expected time for that event and is denoted by T_E.

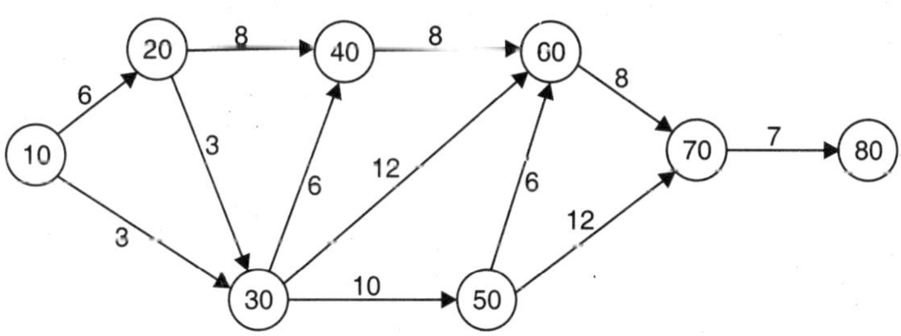

Fig. 7.14

To illustrate the procedure, let us consider the network shown in Fig. 7.14.

$T_E^{10} = 0$

$T_E^{20} = T_E^{10} + T_E^{10-20} = 0 + 6 = 6$

$T_E^{30} = \text{Maximum of } [(T_E^{10} + t_E^{10-30}), (T_E^{20} + t_E^{20-30})]$

$= \text{Maximum of } [(0 + 8), (6 + 3)] = 9$

$T_E^{40} = \text{Maximum of } [(T_E^{20} + t_E^{20-40}), (T_E^{30} + t_E^{30-40})]$

$= \text{Maximum of } [(6 + 8), (9 + 6)] = 15$

$T_E^{50} = T_E^{30} + T_E^{30-50}$

$= 9 + 10 = 19.$

T_E^{60} = Maximum of $(T_E^{30} + t_E^{30-60})$, $(T_E^{40} + t_E^{40-60})$ and $(T_E^{50} + t_E^{50-60})]$

= Maximum of [(9 + 12), (15 + 8) add (19 + 6)] = 25.

Similarly,

T_E^{70} = 33, and

T_E^{80} = 40.

All these computations can be carried on the network itself.

***(b)* Backward Pass Computations:** Calculations begins from the end node of the project and proceed towards the start node. To start the calculations, the time of occurrence for the last node is decided. This is the time at which the project must be completed. This is called the *'Contractual Obligation Time'* and is denoted by T_s. If not known, the contractual obligation time is taken to be equal to the earliest expected time for the end event. The objective of the backward pass is to calculate the *'Latest Allowable Occurrence Time'*, the time at which a particular event must occur at the latest. This is denoted by T_L.

Referring to the the network shown in Fig. 7.14.

$T_L^{80} = T_L^{80} = 40$

$T_L^{70} = T_L^{80} - t_E^{70-80} = 40 - 7 = 33$

$T_L^{60} = T_L^{70} - t_E^{60-70} = 33 - 8 = 25$

T_L^{50} = Minimum of $[(T_L^{60} - t_E^{50-60}), (T_L^{70} - t_E^{50-70})]$

= Minimum of [(25 – 6), (33 – 12)] = 19

$T_L^{40} = T_L^{60} - T_E^{40-60} = 25 - 8 = 17$,

T_L^{30} = Minimum of $[(T_L^{40} - t_E^{30-40}), (T_L^{60} - t_E^{30-60})$ and $(T_L^{50} - t_E^{30-50})]$

= Minimum of [(17 – 6), (25 – 12) and (19 – 10)] = 9.

Similarly, T_L^{20} = 6 and T_L^{10} = 0.

All these computations can be carried on the network itself. The network with the T_E and T_L values for all events is shown in Fig. 7.15.

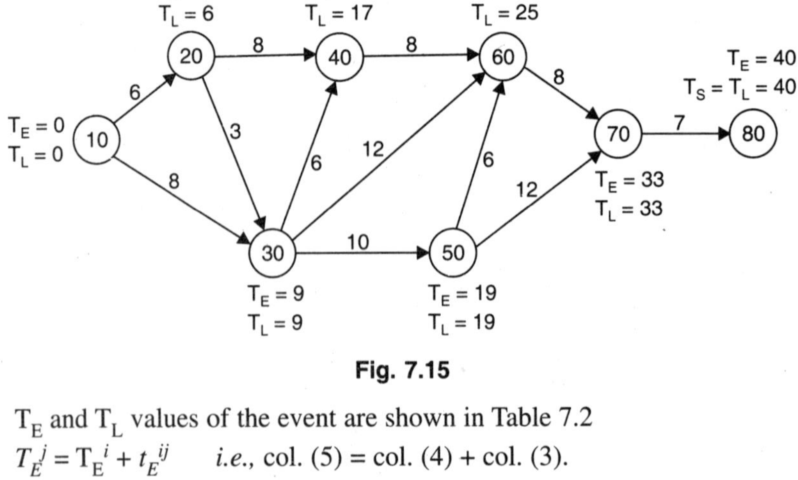

Fig. 7.15

T_E and T_L values of the event are shown in Table 7.2

$T_E^j = T_E^i + t_E^{ij}$ *i.e.*, col. (5) = col. (4) + col. (3).

$T_E^i = T_L^j - t_E^{ij}$ *i.e.*, col. (6) = col. (7) + col. (3).

Table 7.2

Activity ij		*Activity*	*Earliest*		*Latest*	
Predecessor event (i) (1)	*Successor event (j)* (2)	*Duration* (t_E^{ij}) (3)	*Start time* (T_E^i) (4)	*Finish time* (T_E^j) (5)	*Start time* (T_L^i) (6)	*Finish time* (T_L^j) (7)
10	20	6	0	6	0	6
10	30	8	0	8	1	9
20	30	3	6	9	6	9
20	40	8	6	14	9	17
30	40	6	9	15	11	17
30	50	10	9	19	9	19
30	60	12	9	21	13	25
40	60	8	15	23	17	25
50	60	6	19	25	19	25
50	70	12	19	31	21	33
60	70	8	25	33	25	33
70	80	7	33	40	33	40

Slack

Slack and float both refer to the amount of time by which a particular event or activity can be delayed without affecting the time schedule of the network. The term slack refers to events and is used with PERT networks, while the term float refers to activities and is used with CPM network.

In PERT language, slack is defined as the difference between the latest allowable time and the earliest expected time of an event.

For an event j, slack $S_j = T_L^j - T_E^j$.

Table 7.3

Activity ij		*Activity*	*Earliest*		*Latest*		*Slack*
Predecessor event (i) (1)	*Successor event (j)* (2)	*Duration* (t_E^{ij}) (3)	*Start time* (T_E^i) (4)	*Finish time* (T_E^j) (5)	*Start time* (T_L^i) (6)	*Finish time* (T_L^j) (7)	$S_j = T_L^j - T_E^j$ ($= T_L^i - T_E^i$) (8)
4	8	4	0	4	0	4	0
8	12	6	4	10	8	14	4
8	16	8	4	12	4	12	0
12	20	5	10	15	14	19	4
16	20	7	12	19	12	19	0
20	14	5	19	24	19	24	0

Calculation of slacks for the network shown in Fig. 7.16 is given in Table 7.3.

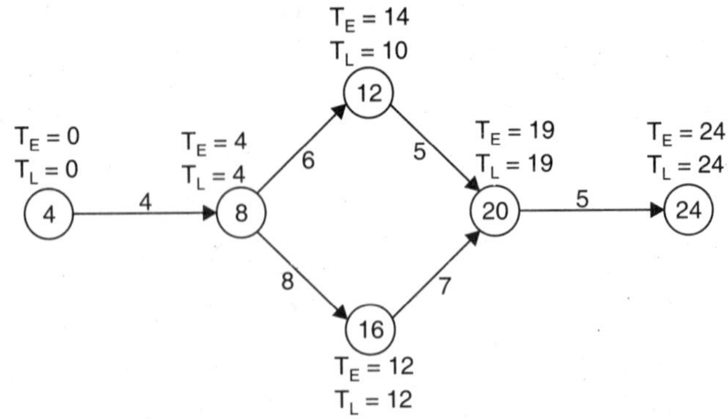

Fig. 7.16

Critical Path

After knowing the T_E and T_L values for various events in the network, the critical path activities can be identified by applying the following conditions:

(*i*) T_E and T_L values for the tail event of the critical activity are the same, *i.e.*, $T_E^i = T_L^i$.

(*ii*) T_E and T_L values for the head events of the critical activities are the same, *i.e.*, $T_E^j = T_L^j$.

Or we can say that both the predecessor and successor events of the critical activity have zero slacks.

(*iii*) For the critical activity, $T_E^j - T_E^i = T_L^i - T_L^i = t_E^{ij}$.

The path forming an unbroken chain of critical activities from the start event to the end event is called the *critical path.* On the critical path all events have zero slacks. In the network the critical path is shown by thick lines.

When we apply the above conditions to the network shown in Fig. 7.17 we find that activities 4–8, 8–16, 16–20 and 20–24 are critical activities. The path 4–8–16–20–24 is, thus, the critical path.

In the execution of the project, it is the critical path which needs optimum care. Any delay in any activity on this path will upset the whole project and will delay its completion. That is why the name critical path is assigned to this path.

Example 7.3: *Consider the network shown in Fig. 7.17. the activity times in days are given along the arrows. Calculate the slacks for the events and determine the critical path. Put the calculations in tabular form.*

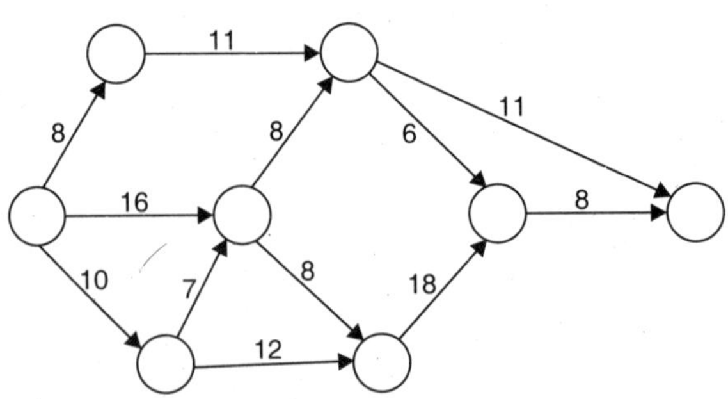

Fig. 7.17

Solution: The events of the network are first numbered and then T_E and T_L values are calculated. Figure 7.18 shows the network with the numbers in the node circles and T_E, T_L and S values along the nodes.

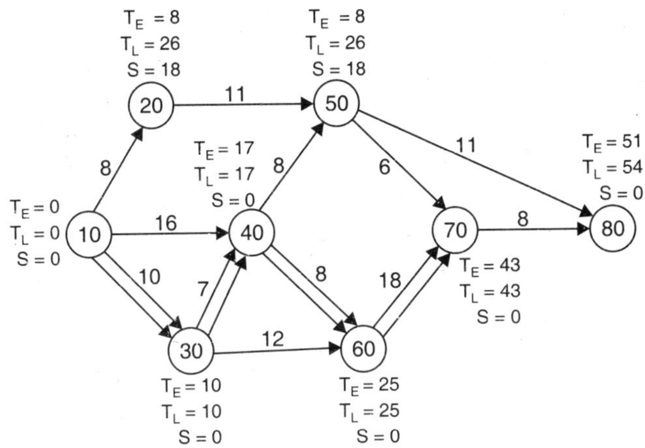

Fig. 7.18

Table 7.4

Activity ij		*Activity*	*Earliest*		*Latest*		*Slack*
Predecessor event (i)	*Successor event (j)*	*Duration* (t_E^{ij})	*Start time* (T_E^i)	*Finish time* (T_E^j)	*Start time* (T_L^i)	*Finish time* (T_L^j)	$S_j = T_L^j - T_E^j$ $(= T_L^i - T_E^i)$
(1)	(2)	(3)	(4)	(5)	(6)	(7)	(8)
10	20	8	0	8	18	26	18
10	30	10	0	10	0	10	0
10	40	16	0	16	1	17	1
20	50	11	8	19	26	37	18
30	40	7	10	17	10	17	0
30	60	12	10	22	13	25	3
40	50	8	17	25	29	37	12
40	60	8	17	25	17	25	0
50	70	6	25	31	37	43	12
50	80	11	25	36	40	51	15
60	70	18	25	43	25	43	0
70	80	8	43	51	43	51	0

From the computations carried on the network as well as in Table 7.4, we find that the slack is zero for activities 10–30, 30–40, 40–60, 60–70, 70–80 and non-zero for all others.

Probability of Achieving Completion Data

The distribution of times associated with activities in a production process is found to be a β-distribution curves. If an activity is having three time estimates the t_o, t_m and t_p, then the expected time to complete the activity.

$$t_e = \frac{t_o + 4t_m + t_p}{6}$$

Standard deviation $\sigma = \frac{t_p - t_o}{6}$

and variance $\sigma^2 = \left(\frac{t_p - t_o}{6}\right)^2$

By use of factor z, derived from the standard deviation, expected time and scheduled time, and by reference to table produced from the β-distribution curve, it is possible to ascertain the probability of an activity being completed by the scheduled time. Here, we assume that the distribution for the total duration of the project is normal.

Then the standard normal variate for the desired time (scheduled time) t_s is given by

$$z = \frac{\text{Scheduled time} - \text{Expected time}}{\text{Standard deviation}}$$

i.e., $z = \frac{t_s - t_e}{\sigma}$

From the table corresponding to the computed value of z, the probability of achieving the scheduled time can be determined.

If the scheduled time is made the same as the expected time, $z = 0$, the probability for this value of z is 50 percent. In other words, there is a fifty-fifty chance of the scheduled data being achieved. Generally, 65 percent probability is accepted as providing sufficient certainty associated with the economical use of resources. A higher degree of certainty is only sought when other reasons override economic consideration.

For a project consisting of number of activities, to calculate the probability of achieving the completion data, it is essential to compute 't' and 'σ' for each activity. Then the total expected time is the arithmetic sum of individual 't_e' for each activity, i.e., $\Sigma t_e = t_{e1} + t_{e2} + \ldots$

The standard deviation for the last event (network ending event) in a series of activities, is given by $\sigma = \sqrt{\sigma_1^2 + \sigma_2^2 + \sigma_3^2 + \ldots\ldots \sigma_n^2}$

Table 7.5 Standard Normal Distribution Function

Z(+)	*Probability (Pr)(%)*	Z(–)	*Probability (Pr)(%)*
0	50.0	0	50.0
+0.1	53.98	–0.1	46.02
+0.2	57.93	–0.2	42.07
+0.3	61.79	–0.3	38.21
+0.4	65.54	–0.4	34.46
+0.5	69.15	–0.5	30.85
+0.6	72.57	–0.6	27.43
+0.7	75.80	–0.7	24.20
+0.8	78.81	–0.8	21.19

+0.9	81.59	–0.9	18.41
+1.0	84.13	–1.0	15.87
+1.1	86.43	–1.1	13.57
+1.2	88.49	–1.2	11.51
+1.3	90.32	–1.3	9.68
+1.4	91.92	–1.4	8.08
+1.5	93.32	–1.5	6.68
+1.6	94.52	–1.6	5.48
+1.7	95.54	–1.7	4.46
+1.8	96.41	–1.8	3.59
+1.9	97.13	–1.9	2.87
+2.0	97.72	–2.0	2.28
+2.1	98.21	–2.1	1.79
+2.2	98.61	–2.2	1.39
+2.3	98.93	–2.3	1.07
+2.4	99.18	–2.4	0.82
+2.5	99.38	–2.5	0.62
+2.6	99.53	–2.6	0.47
+2.7	99.65	–2.7	0.35
+2.8	99.74	–2.8	0.26
+2.9	99.81	–2.9	9.19
+3.0	99.87	–3.0	0.13

Example 7.4: *A project isr expected to take 12 months along the critical path, having a standard deviation of 4 months. What is the probability of completing the project within (a) 10 months (b) 16 months (c) 8 months.*

Solution: (*a*) The probability factor is given by

$$z = \frac{T_S - T_E}{4} = \frac{10 \quad 12}{4} = -0.5$$

From table for $z = -0.5$, probability = 30.85%

(*b*) For $T_S = 16$ months

$$z = \frac{16 - 12}{4} = 1$$

From table for $z = 1$, probability = 84.13%

(*c*) For $T_S = 8$ months

$$z = \frac{8 - 12}{4} = -1$$

Probability for $z = -1$ is 15.87%.

CPM Computations

CPM emphasis more on activities and PERT emphasis more on events. CPM does not take into account the uncertainties while estimating its activity times.

CPM Terms

Earlier Start Time: It is the earliest possible time at which an activity can start, and is calculated by moving from first to last event in the network diagram.

$$E_S^{\,l} = T_E^{\,l}.$$

Earliest Finish Time: This is the earliest start time of the activity plus the duration of the activity.

$$E_C^{lj} = E_S^{l} + t_E^{lj} = T_E^{l} + t_E^{lj}.$$

Latest Start Time: This is the latest possible time by which an activity can start without delaying the date of completion of project.

$$L_S^{lj} = L_C^{j} - t_E^{lj} = T_L^{j} - t_E^{lj}.$$

Latest Finish Time: It is the latest time by which the activity must be completed so that the scheduled date for the completion of the project may not be delayed.

$$L_C^{j} = T_L^{j}.$$

Float: Float of the activity has the same significance as the slack of the events. Slack corresponds to events hence to PERT while the float corresponds to activities hence to CPM.

$$\text{Float} = (\text{LST–EST}) \text{ or } (\text{LFT–EFT})$$

$$(L_S^{lj} - E_S^{l}) \text{ or } (L_C^{j} - E_C^{lj})$$

Exarnple 7.5: *Consider the network shown in Fig. 7.19. Determine the float and identify the critical path.*

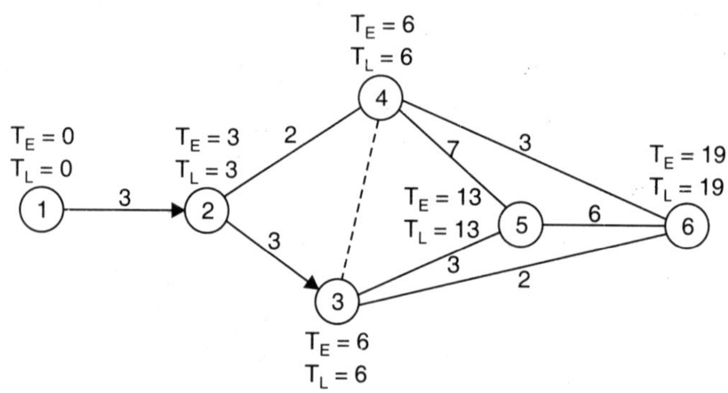

Fig. 7.19

Solution:

Table 7.6

Activity ij	*Activity duration* t_E^{ij}	*Earliest*		*Latest*		*Float*
		(T_{E1}) *Start time* E_S	(T_{E2}) *Completion time* (E_C^{ij})	(T_{L1}) *Start time* (L_S^{ij})	(T_{L2}) *Completion time* (T_L^{i})	
1	**2**	**3**	**4**	**5**	**6**	**7**
1—2	3	0	3	0	3	0
2—3	3	3	6	3	6	0
2—4	2	3	5	4	6	1
3—4	0	6	6	6	6	0
3—5	3	6	9	10	13	4
3—6	2	6	8	17	19	11
4—5	7	6	13	6	13	0
4—6	5	6	11	14	19	8
5—6	6	13	19	13	19	0

The critical path is 1–2–3–4–5–6.

Crashing of Network: Sometimes it is desired to accomplish the project in shorter duration of time. In order to avoid penalties because of late completion of the project. The crashing of the network involves considerations of the cost incurred on the various activities. The following terms need to be defined in connection with the crashing of the network.

(*a*) *Normal Cost:* It is the lowest cost of completing an activity in the minimum time, employing normal means *i.e.*, without using over time or special resources.

(*b*) *Normal Time:* It is the minimum time required to achieve the normal cost.

(*c*) *Crash Cost:* It is the least cost of completing an activity by employing all possible means like over time, additional resources in the form of machines, equipment and materials.

(*d*) *Crash Time:* It is the absolute minimum time associated with the crash cost.

The cost-time slope of an activity is determined as:

$$S = \frac{(\text{Crash Cost} - \text{Normal Cost})}{(\text{Normal Time} - \text{Crash Time})}$$

The slope *S* determines the maximum compression of each activity which is the difference between the normal time and the crash time. It is shown in Fig. 7.20.

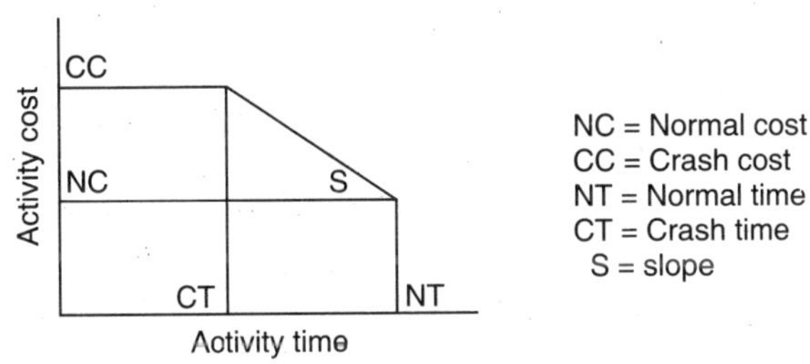

Fig. 7.20

Smoothing: Smoothing reters to scheduling the activities within the limits of their total floats such that fluctuations in the resource requirements arc minimized. The smoothing technique makes use of floats in different activities. The activities are adjusted within a period equal to their duration plus float. In other words an activity having float may be started at any time during this period and finished within the same period. The activity can be taken during the earlier, middle or latter part of the total period consisting of duration plus float. The technique of resource smoothing can be illustrated by an example.

Figure 7.21 shows an arrow diagram with activity duration and number of workers required to complete each activity. The initial schedule graph in which various activities have been placed with the number of workers written on them are shown in Figure 7.22. Figure 7.23 shows the manpower of loading chart and work load distribution chart from the data of Figure 7.22. This chart shows that the number of workers required on any day are four and maximum work forqe size is 15. There is a large variation in the day-to-day load distribution. Figure 7.24 shows the modified schedule graph in which activities 2–5 and 5–8 have been shifted to extreme right hand side and activities 1–3, 3–6 and 6–8 have been displaced from extreme left towards the middle of the graph. Activities 1–4, 4–7 and 7–8 have no float and hence cannot be shifted. Modified manpower loading chart as shown in Fig. 7.25 is an outcome of modified schedule graph and shows better smoothing as compare to Fig. 7.23. The variation of work force is 7 (14–7) against 11 (15–4) in case of initial manpower loading chart. Fig. 7.24 is further modified to Fig. 7.26 by reshifting the activities 1–3, 3–6 and 6–8 and Fig. 7.27 shows the further modified manpower loading chart which gives work force variation of only 4 (11–7) and is still better smoothing of the work load. The ideal manpower loading chart as shown in Fig. 7.27 can rarely be achieved in normal real life projects but one should always aim to approach this ideal manpower loading chart.

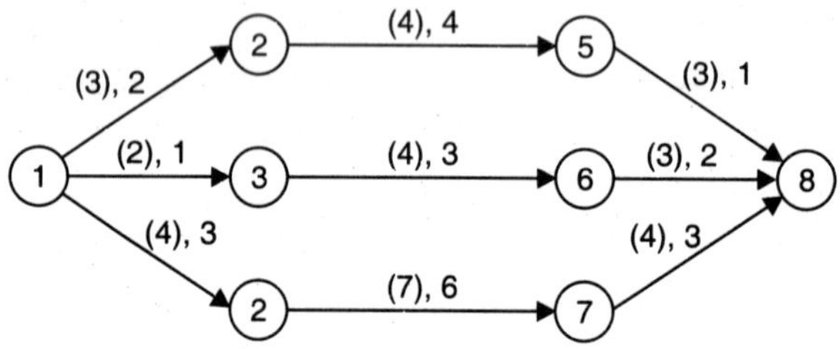

Fig. 7.21 Arrow diagram.

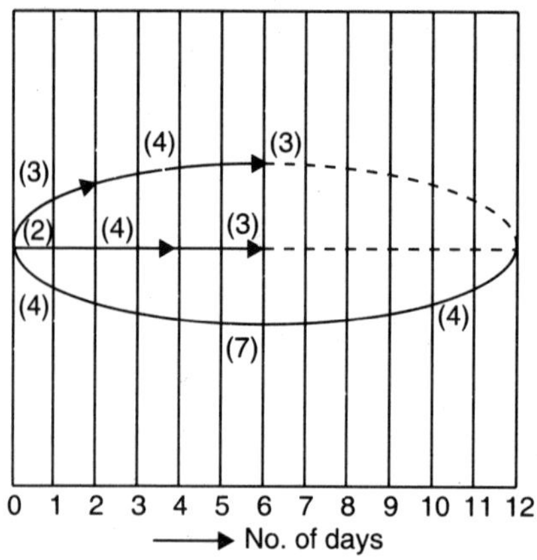

Fig. 7.22 Initial Schedule graph.

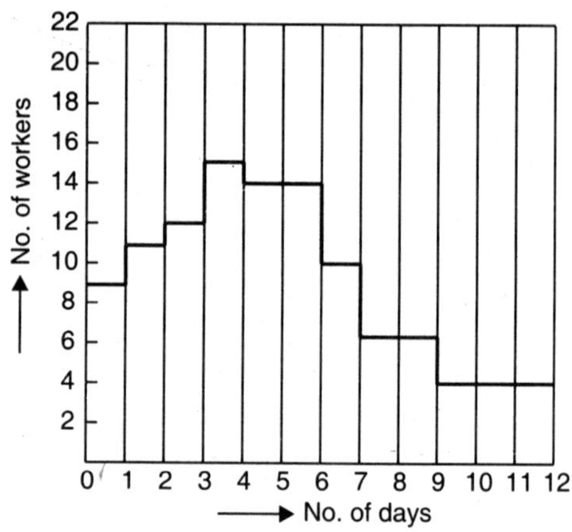

Fig. 7.23 Manpower Loading Chart.

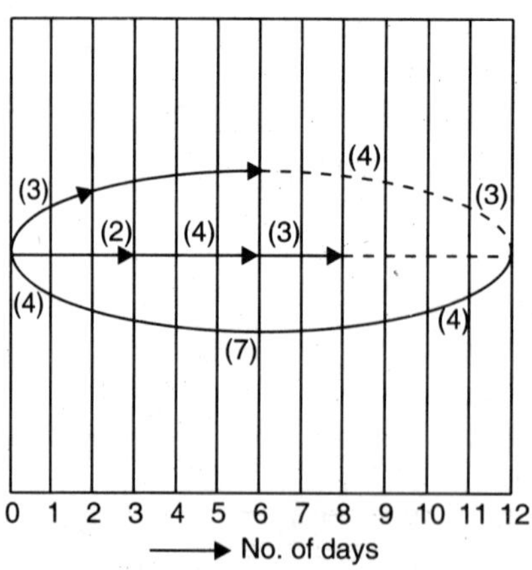

Fig. 7.24 Modified schedule graph.

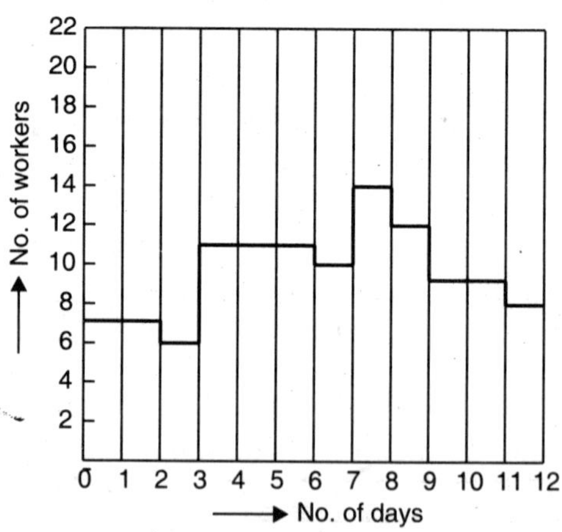

Fig. 7.25 Modified manpower loading chart.

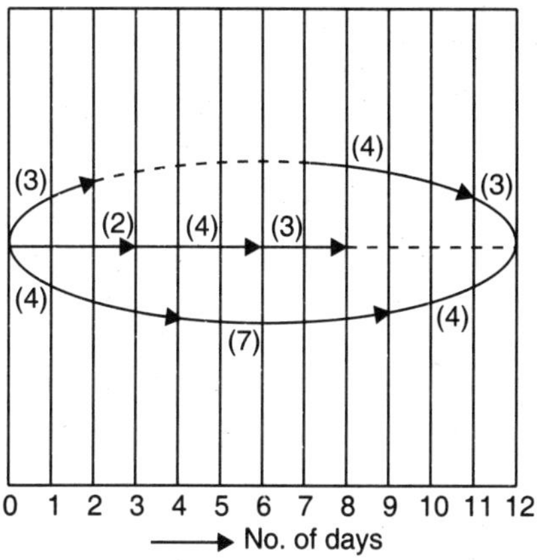

Fig. 7.26 Further modified schedule grdaph.

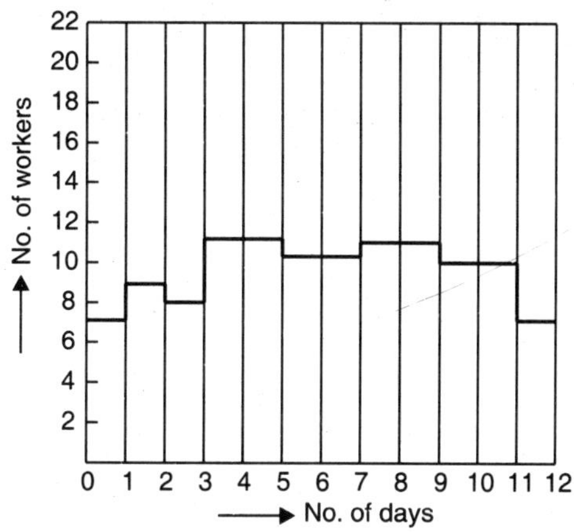

Fig. 7.27 Further modified manpower loading char.

Applications oFNetwork Teeftniques

Network techniques have been used with advantage for solution of problems of diverse nature in various fields of activity. They serve as versatile tools for planning and scheduling of variety of jobs. Some of the typical applications of network technique are as follows:

- Planning and development of space programmes.
- Development of new weapon systems.
- Development and launching of new products.
- For construction of buildings, roads, bridges, railways, airports, etc.
- For maintenance planning.
- For installation of heavy machinery.
- For construction of thermal power plants.
- For setting up chemical plants, oil refineries, fertilizer plants, etc.
- For planning seminars and conferences.
- For production control.

7.4 PROJECT MONITORING

Project monitoring may be defined as the process of inducing action for adherence to project schedule. It involves three steps—measuring, reviewing and reporting.

Measuring: Measuring process first measures what has been done in relation to what should have been done? This information can either be gathered by monitoring agency directly through observing the project work continuously or could be provided by the executing agencies responsible for project work on specified formats provided by the monitoring agencies.

Reviewing: The information obtained is processed for computing an index known as percentage progress. The percentage progress is a composite index which enables evaluation of the achievement vis-a-vis commitment for actual physical progress; physical progress with reference to time, cost and effort; and labour time, and capital productivity.

Reporting: The gap between commitment and achievement is then reported to the concerned executives for taking necessary actions.

Project-Monitoring Techniques

Techniques used for monitoring the projects relate to progress measurements and performance monitoring. The two specific techniques used for progress measurement and performance measurement namely Line of Balance and PACE are discussed here.

Line of Balance: Line of Balance is a managerial technique to ascertain the progress of various phases of a project and to maintain a balance by applying the concept of management by exception. The technique was developed by U.S. Navy during World War II to maintain control over the various production contracts. The technique monitors the progress of the various phases of the project at a given point of time by determining, (*i*) the desired progress of all phases of progress of the various phases and (*ii*) the phases which are not making desired progress.

The project manager then investigates the cause of delay in these phases and takes corrective actions so that the project completion schedule may not be delayed. The line of balance technique comprises of four elements, *viz.* (*i*) an objective chart, (*ii*) a plan of operations, (*iii*) a progress chart and (*iv*) analysis of the progress chart.

The objective chart contains scheduled curves for each phase of the project and one for the entire project. The plans of operations depicts each phase within a project with starting and ending points placed on project plans with the established time for the phase separating the two points. The phases are laid out on the plans so as to complete the project in a stipulated time with proper utilization of facilities and considering all relative factors. Each phase is depicted with the help of a symbol. The starting symbol is left blank and the ending symbol is cross-hatched.

The progress chart is drawn in the form of vertical columns representing the phases of the project and one column representing the total project. The vertical scale represents the percentage completion and is of the same scale as in the objective chart.

Line of balance is drawn directly from the objective chart, with a vertical line drawn representing the date of study. The intersection point of date of study with the scheduled curves of various phases of the project reports the percentage progress of these phases. The line of balance for the total project is determined from the intersection of the date of study line and the total project schedule curve, and is used to compute the total project percentage completion. The line of balance technique is illustrated by the following example.

A project related to the developing of a prototype for a defense aircraft consists of thirteen phases as given below:

Phase	*Phase No.*	*Total Estimated Months*
Wing Design	1	10
Fuselage Design	2	11
Gunmount Design	3	7
Tail Assembly, Design	4	9
Landing Gear Design	5	8
Wing Manufacture	6	7

Fuselage Manufacture	7	9
Gunmount Manufacture	8	5
Landing Gear Manufacture	9	6
Tail Assembly Manufacture	10	5
Wing-Gunmount Assembly	11	2
Airframe Assembly	12	5
Final Assembly	13	3
Overall Project		87

The project consists of five design phases and five manufacturing phases and three assembly phases. The plan of operations, the objective chart and the progress chart for this project have been shown in Figure 7.28, 7.29 and 7.30 respectively.

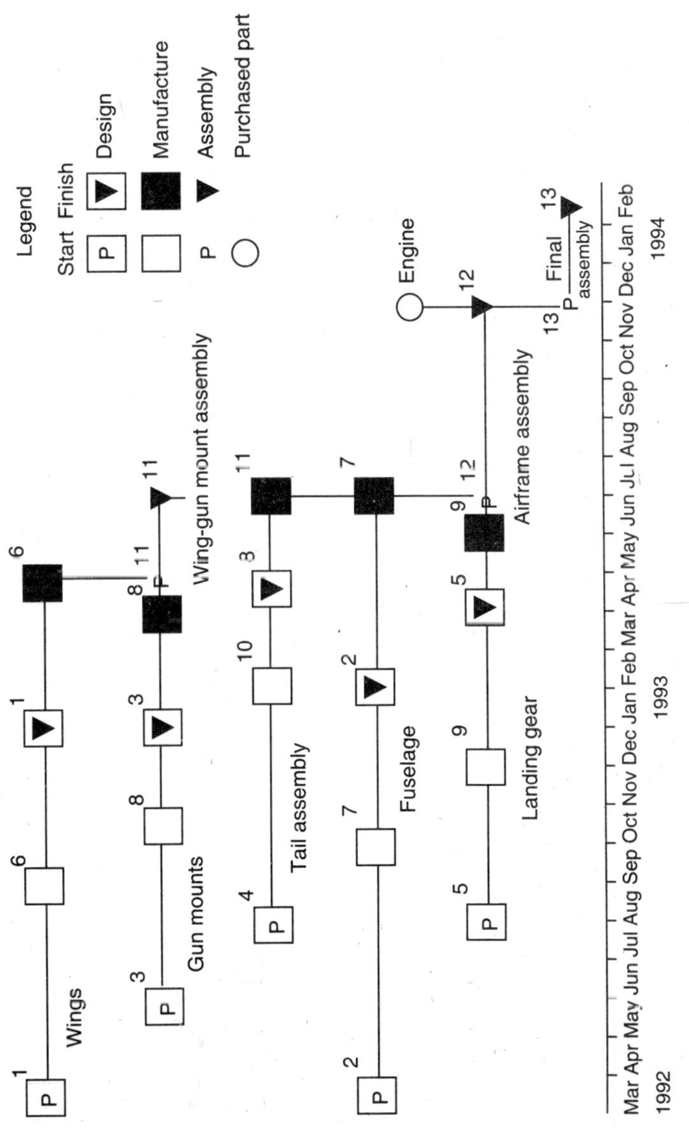

Fig. 7.28 Plan of operations for research and development project.

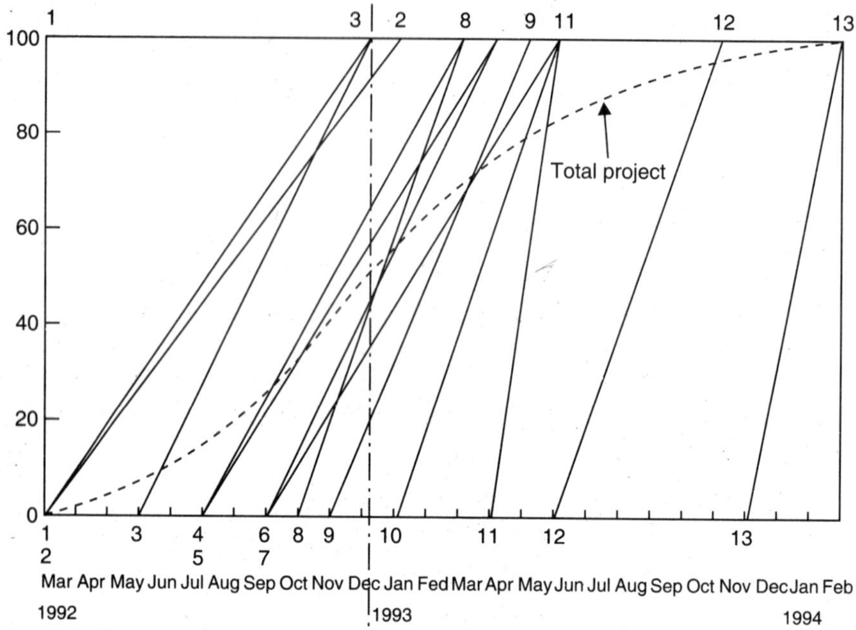

Fig. 7.29 Objective chart for research and development project.

The Plan of Operations: The Plan of operations is shown in Fig. 7.28. The project is planned to cover 24 months starting from March 1992 and is scheduled to finish by February 1994. Each phase has a starting and an ending point placed on the plan with the standard months required for the phase separating the two points. The phases are laid out on the plan so as to lead into the final product. Each phase is depicted by a special symbol to facilitate the reading of the plan.

The Objective Chart. The objective chart is depicted in Fig. 7.30. It contains a scheduled curve for each phase and one for the entire project. In the given example there are thirteen phases, the objective chart therefore contains fourteen schedule curves,

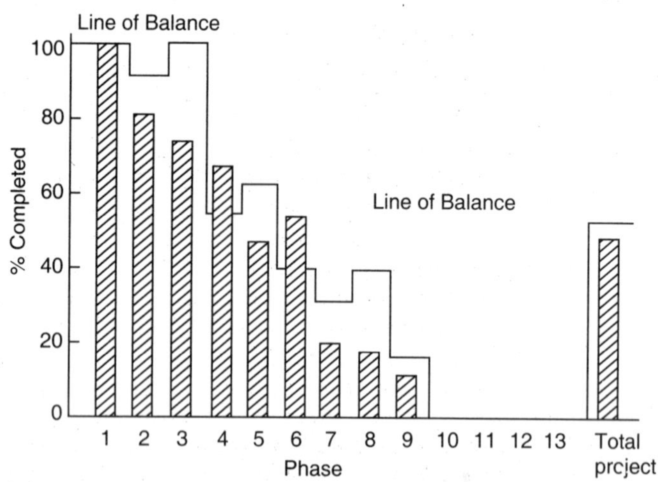

Fig. 7.30 Objective chart for research and development project.

one for each phase and one for the entire project. The schedule curves for project phases are shown as straight lines. The vertical scale of the objective chart represent the percentage of completion. The starting point of phase one is March 1, 1992. The completion date is January 1, 1993. A straight line is drawn from 0% completed at March 1, 1992, to 100% completed at January 1, 1993. The curve is labelled 1 at each end. Each of the other twelve phases schedules are drawn similarly.

The total project schedule curve is derived from the cumulative schedule of phase months shown in Table 7.7.

Table 7.7

	Month	*Phase Months Planned/Month*	*Cumulative Phase Months*	*Cumulative % Phase Months*
1992	March	2	2	2.30
	April	2	4	4.60
	May	2	6	6.90
	June	3	9	10.34
	July	3	12	13.79
	Aug	5	17	19.54
	Sep	5	22	22.59
	Oct	7	29	33.33
	Nov	8	37	42.53
	Dec	9	46	52.87
1993	Jan	7	53	60.92
	Feb	7	60	68.97
	March	7	67	77.01
	April	5	72	82.96
	May	4	76	87.36
	June	3	79	90.80
	July	1	80	91.95
	Aug	1	81	93.10
	Sep	1	82	94.25
	Oct	1	83	95.40
	Nov	1	84	96.55
	Dec	1	85	97.70
1994	Jan	1	86	98.85
	Feb	1	87	100.00
		87		

Column 2 of the above table is obtained from the plan of operations shown in Fig. 7.28. During the month of March 1992, two phase months were in operation, one for phase 1 and one for phase 2. During the month of Dec. 1992, 9 phase months were in operation-phases 1, 6, 3, 8, 4, 2, 7, 5 and 9. The phase months are accumulated in column 3 of the Table and expressed as a cumulative percentage in column 4. The cumulative percentage is then grouped on the objective chart as total project schedule curve as shown by the dotted curve.

Progress Chart: Vertical column represents the phase of the project with one column representing the total project. The vertical scale is in terms of completion. The line of balance is derived directly from the objective chart, on which has been drawn a vertical line representing a date of study, in this example Jan. 1, 1993. The data of study line intersects the schedule curve for phase 1 at the 100% level, the line of balance for phase 1 is therefore drawn at 100% on the progress chart. The intersection for phase 2 is at 91%, for phase 3 at 100% and so on. As the starting points of phases 10 to 13 fall after the date of study, the line of balance of these phases drops to the 0% level. The height of the vertical bars on the progress chart is derived from the following Table 7.8, depicting the analysis of phase time requirements. Column 3 represents an estimate of the months required to complete each phase as of the date of study. These estimates are made independent of the original time estimates and should be realistic in the light of unforeseen difficulties or any other considerations which may have evolved since the original estimates were made. Column 4 represents the percentage of completion in terms of time. This is derived by subtracting the figure in column 3 from that of column 2 and dividing the difference by the Fig. in column 2. The quotients is then multiplied by 100 to convert it to a percentage.

Table 7.8

Phase No.	*Original Estimated Months*	*Estimated Months as on date of Study*	*Estimated % completion*
1	10	0	100
2	11	2	82
3	7	2	71
4	9	3	66
5	8	4	50
6	7	3	57
7	9	7	22
8	5	4	20
9	6	5.5	8
10	5	5	0
11	2	2	0
12	5	4	0
13	3	3	0
	87	44.5	

The percentage phase completed is given as = $100\,(b - a)/(b)$

where a = estimated months to complete the phase

b = original estimated months required for the entire phase

Since phase 1 is completed, the vertical bar represents 100% level. The completion of phase 2 is given by

% phase completed = 100 (11 – 2)/(11) = 82%

To determine the height of vertical bar for the total project the original total phase months for entire project is obtained by totaling column 2 and total of estimated months to be completed is obtained totalling column 3. The percentage completion is then determined by

Total % completion = 100 $(B - A)/(B)$

where A = Total original estimated phase months for the entire project

B = Total estimated phase months required to complete the project as of the date of study.

The completion for the total project for the given example:

Total project % completed = 100 (87 – 44.5)/(87) = 49%

The line of balance for the total project is derived from the intersection of the date of study line and the total project schedule curve. In the given example the intersection is at 53%.

Analysis of the Progress Chart: The analysis of the progress chart reveals the conditions of the entire project at a glance. Three of the five design phases are behind schedule. These behind scheduled conditions have forced the related manufacturing phases also to be behind schedule. The tail assembly design schedule is well ahead of the schedule. A possible action may be to transfer some of the design engineers from tail assembly group to behind-schedule fuselage design, gunmount design and landing gear design groups. However, it may not be feasible because of the different skills required by different groups. Alternative actions may have to be thought for making the desired progress in these phases.

PACE (Performance And Cost Evaluation): PACE is statistical analytical method to monitor the performance of a group of people working on a project, through monitoring of PACE measurement index computed as follows:

$$\text{PACE Index} = \frac{[\text{N} \pm \text{L} - (\text{I} + \text{A})]}{\text{N} \pm \text{L}} \times \text{E} \times 100\%$$

where N = Number of persons assigned on the project.

L = Number of persons temporarily transferred to and from the project work (persons on loan). Positive sign indicates persons taken on loan for project work and negative sign indicates persons given away from the project work.

I = Number of persons which are not specifically working on project work (Idle persons).

A = Number of persons out of project area (all absences from the project work except those when the employees are authorized to perform the project work at some other location).

The related indices are as follows:

$$\text{Idle time Index} = \frac{\text{Average No. of Idle persons}}{\text{Number assigned + Persons on loan}} \times 100\%$$

$$\text{Out of Area Index} = 100 - \frac{\text{No. of workers observed}}{\text{No. assigned + Persons on loan}} \times 100\%$$

$$\text{Group Effort Rating} = \frac{\text{Sum of rating (\%) of the individual workers}}{\text{No. of persons in the group}} \times 100\%$$

Example 7.6: *The following observations are made deriving the persons working on project for first 11 weeks.*

Week	*No. of persons assigned*	*No. of persons loaned*	*Average No. Idle*	*% idle*	*Average No. observed*	*% Out of Area*	*% Effort*	*% PACE*
1	120	0	12	10	110	8	80	65
2	120	10	42	32	117	10	83	48
3	120	11	45	34	122	7	79	46
4	120	8	9	7	123	4	70	62
5	120	7	15	12	119	6	70	57
6	120	2	7	6	118	3	75	68
7	120	0	8	7	118	2	80	73
8	120	–4	6	5	114	2	76	71
9	116	–6	3	3	106	4	77	72
10	110	–8	5	5	99	3	85	78
11	102	–3	8	8	95	3	86	77

Solution: Sample Calculations

PACE index for First week:

$$N = 120$$
$$L = 0$$
$$I = 12$$
$$A = 120 - 110 = 10$$

E = 80% (It represents the average rating of persons working during the first week).

Therefore,

$$\text{PACE Index} = \frac{120 + 0 - (12 + 10)}{120 + 0} \times 80 \times 100 = 65\%$$

$$\text{Idle Time Index} = \frac{\text{Average No. Idle}}{\text{No. Assigned + Loans}} \times 100 = \frac{12}{120 + 0} \times 100 = 10$$

$$\text{Out of Area Index} = 100 - \frac{\text{No. of workers observed}}{\text{No. Assigned + Loans}} \times 100$$

$$= 100 - \frac{110}{120 + 0} \times 100 = 8\%$$

The PACE index along with its components, *e.g.* idle time index and out of area index are shown in Fig. 7.31 for the first 11 weeks of the industrial project given in the above example. The lower 5% shaded portion in Fig. 7.31 indicates that the objective of PACE chart is to monitor the idle time index and out of area index to within this area. An idle time index and out of area index within the shaded area (5% or less) are regarded as being in control. If during a week these values fall above the shaded area, there is need to investigate the possible causes.

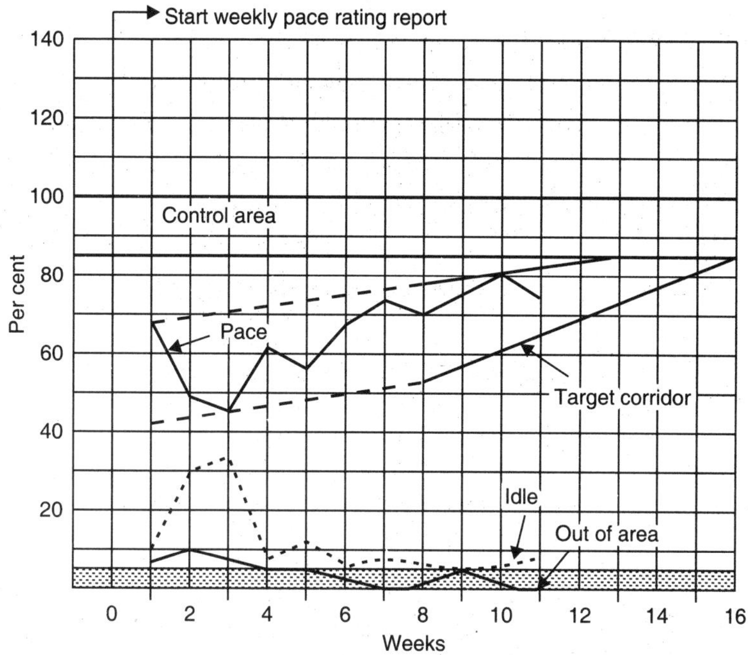

Fig. 7.31 PACE measurement index and components.

The PACE measurement index is considered in control when it is in the 85% to 100% range as shown in Fig. 7.31. The 85% lower range is derived from a 5% error allowance for each of the three components of the PACE index namely, idle time, out of area and effort rating. Few projects will have an initial PACE index in the control area during the initial few weeks of implementing the project. The PACE index tend to move towards the control area after a few weeks of efforts made by project manager/ coordinator to improve upon the excessive idle time, out of area index, and too low group effort index. After about 8 weeks, the upward trend becomes evident and an upper and lower control lines are drawn on the PACE chart leading into the control area. These control lines should follow the trend so far obtained, leading into the control area at around 15 to 16 weeks after the project is implemented. The area between the control lines is called the target corridor. PACE readings outside the target corridor are investigated. Those far below are investigated so that improvements could be made fast enough and readings above the upper control are investigated to find how the exceptional performance was achieved, so as to continue the same.

Project Control

Project control involves a regular comparison of performance against targets, a search for causes of deviation, and a commitment to check adverse variances. Effective project control is essential for realizing the project objectives by, (*i*) ensuring regular monitoring of the performance, (*ii*) motivating project personnel to strive for achieving project objectives.

Elements of Project Control System

Project Objective: The project objectives(s) need to be stated clearly for designing an effective project control system.

Project Planning: It relates to identification of resources such as manpower, money, material etc. together with the constraints on the availability and devising a scheme of operations for realizing project objectives by making optimum use of available resources.

Project Scheduling: Project scheduling involves listing of various activities related to the project work, sequencing the activities, fixing the time frame for the start and completion of these activities and firm commitment of resources required date by date for completion of these activities.

Project Implementation: Project scheduling provides a basis for implementing project activities by the authorized project personnel/agencies. It leads to generating progress reports useful for project monitoring. A progress report is account of the status of activities under implementation. It is needed by project personnel/agencies responsible for monitoring the project to keep track of the work till its completion.

Project Monitoring: Project monitoring is the process of inducing action for adherence to project schedules based on review of progress reports during implementation of projects.

Project Updating: Project monitoring results in updating of project schedules, incorporating progress information/actual occurrence of events as indicated by project schedule which has been implemented as well as incorporation of new decision/ information pertaining to the schedules yet to be implemented.

Project Control: It refers to overt and covert actions of the project managers to ensure adherence to predetermine project standards related to three key objectives—time, cost and quality. The covert actions refer to the setting of objectives, organization, systems and procedures and the overt actions refer to actions taken as a result of project monitoring.

The factors contributing to the success of a project control system are the commitment of the project personnel to the objectives of the project, delegation of authority to personnel/agencies responsible for project monitoring, careful application of project scheduling and monitoring techniques for project work, continuous updating of project data on sequence, productivity, resource and time utilization during the life span of the project and proper training of the project personnel related to the scheduling and monitoring techniques used.

Problems in Project Control

- Inadequate definition of the project objectives.
- Unjustified changes in the project specifications during the implementation phase of the project.
- Unclear division of responsibilities and deployment of too many agencies (contractors).
- Non-availability of basic data related to project planning and scheduling.
- Over optimistic time schedules.
- Inadequate resource planning, mobilization and distribution.
- Complex and complicated procedures.
- Failure of vendors and sub-contractors to meet commitment.
- Poor communication and coordination.

- Incorrect sequencing of project activities.
- Inadequate design facilities with vendors.
- Lack of essential resources such as skilled labour, consumables, etc.
- Spoilages and defects detected too late.
- Too much emphasis on indigenous development.
- Incompetent implementation authority.
- Power shortages.
- Strikes and lockouts.
- Inadequate construction equipment and tools.
- Lack of transportation facilities.
- Inadequate material handling systems.

EXERCISES

1. Difference between PERT and CPM.
2. Define critical path.
3. Explain crashing of network.
4. Explain different forms of project organisation.
5. Define project scheduling. What are methods of project scheduling?
6. Define project control. What are elements of project control?

❑❑❑

CHAPTER 8

Loading, Scheduling and Sequencing of Operations

8.1 INTRODUCTION

To utilize the facilities of a productive system effeciently and effectively and in order to meet the customer delivery schedules, the jobs to be processed on the various facilities need to be loaded, scheduled and sequenced in an optimum manner. Loading refers to the assigning of jobs to facilities and the committing of facilities to jobs without specifying the time and sequence. Scheduling is the assignment of the work to a facilities to jobs without specifying the time and sequence. Scheduling is the assignment of the work to a facility and the specification of the time and the sequence in which the work is to be done. It is the time phasing of loading. The loading functions are usually carried out at the higher echelons of the organisation, whereas the scheduling function is carried out at the lower echelons. The sequencing of operations refers to a systematic procedure of determining the sequence in which a series of jobs shall be processed on a finite number of service facilities.

8.2 LOADING

Loading involves assigning jobs to work centers and to various machines in the work centres. If a job can be processed on only one machine, no difficulty is presented. However, if a job can be loaded on multiple work centers or machines, and there are multiple jobs to process, the assignment process becomes more complicated. The schedules needs some way to assign jobs to the centers in such a way that processing and setups are minimized along with idle time and throughput time.

8.2.1 Loading Methods

There are two approaches which are used for loading work centres.

1. **Infinite Loading:** It refers to assigning jobs to work centres without considering the capacity of the work center. Jobs are loaded for the production schedule into the facility as if capacity were infinite. Gantt Load Charts and visual load profiles are useful to depict graphically the workload of each facility against a time scale.

(*a*) **Gantt Load Charts:** Gantt load charts graphicalty represent the workloads of facilities on a time scale. Figure 8.1 represents a Gantt load chart for a machine shop having five facilities—lathe, drilling machine, milling machine, grinding machine and testing equipment. Five Job orders are to be processed on these facilities. A copy of every job order is placed in a file which moves with the job going to different facilities. The copy of the job order remains in the file till the job is completed. The job orders placed in the file specify both the facilities used to do each operation and the estimated time to do the operation. The workload for each facility can be prepared with the help of a load analysis sheet given in Table 8.1.

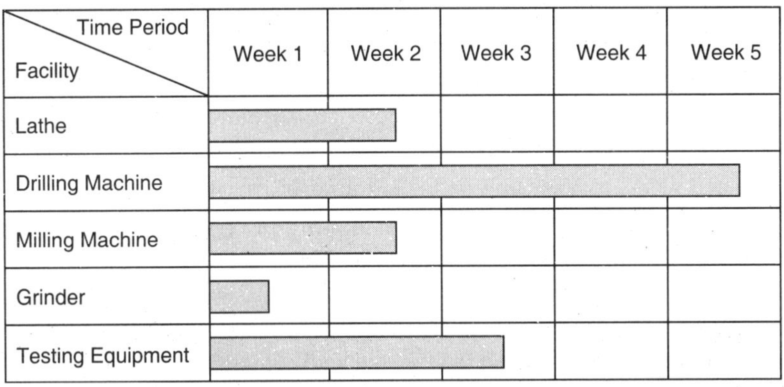

Fig. 8.1 Gantt load chart.

Table 8.1 Load Analysis Sheet

Order No.	*Processing Time on Facilities (Hours)*				
	Lathe	*Drilling M/C*	*Milling M/C*	*Grinder*	*Testing Equip.*
1	15	44	13	4	18
2	18	–	22	3	20
3	–	48	–	5	13
4	16	52	11	–	17
5	11	36	14	8	22
Total Load	60	180	60	20	90

From the load analysis sheet the total number of hours of work for each facility can be determined. For example, assuming that the facilities can be loaded for forty hours, the time available per week—lathe has 60/40 *i.e.*, 1.5 weeks workload ahead of it. Similarly, drilling machine, milling machine, grinder and testing equipment have 4.5, 1.5, 0.5 and 2.25 weeks of workload ahead, respectively. As shown in the Fig. 8.1 the Gantt load chart makes it possible to determine quickly the overall load condition in the shop. For example, the drilling machine has considerably more workload than any other facility. On the other hand, the grinder has an extremely small workload. Both these situations need to be investigated since over load and under load are both undesirable.

Advantages

- The system provides a very simple and easy way to depict the loading of various facilities of a work centre.

- The system can be maintained even by clerical staff when trained to analyze work order file.
- The overall cost of operating the system is relatively small.
- The modification of the system makes it applicable to many situations and various types of work.

Disadvantages

- The information is very gross and therefore can be misleading at times.
- The load chart does not specify the time when each job will be completed, or the sequence of processing the jobs.
- It does not account for variations in human performance and equipment breakdown.
- The chart must be updated periodically to account for new jobs.

This type of loading technique can be successfully used in manufacturing shops, for maintenance work, for research and development activities, etc.

(*b*) **Visual Load Profiles:** A visual load profile is graph comparing workloads and capacities on a time scale. It is shown in the Fig. 8.2. In a manual system the load consists of open orders (existing orders from customers) assigned to the work centre. As shown in the figure, load for week 1 exceeds the capacity but future loads are well within capacity. In a computer-based system, the load consists of open as well as planned orders. As shown in figure, planned loads during weeks 3, 4 and 6 exceed capacity while loads for open orders are feasible, except for week 2.

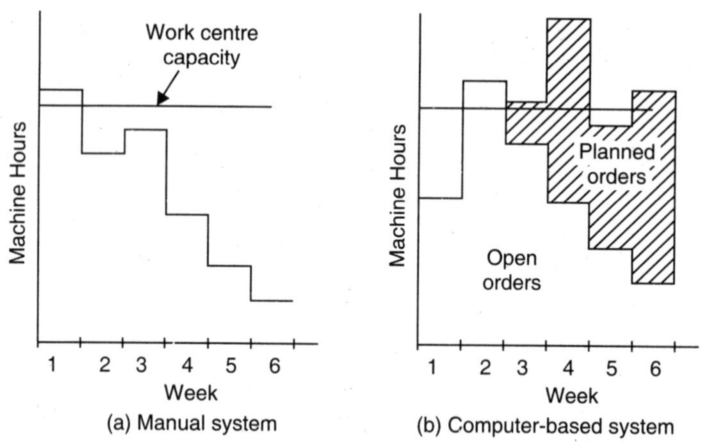

Fig. 8.2 Infinite loading for a work centre—visual load profiles.

2. **Finite Loading:** It starts with a specified capacity of each facility and a list of jobs. The facility's capacity is then allocated unit by unit (*e.g.*, man hour or machine hour) to the jobs by simulating jobs starting times and completion times. Finite loading thus creates a detailed schedule for each job and each facility based on the capacity of the facility. Jobs are allocated to the facilities according to their capacities hour-by-hour, day-by-day, into the future. Figure 8.3 depicts a finite loading system of a work centre having a fixed capacity of 100 machine hours. Finite loading combines the loading, sequencing, the detailed scheduling in a single system. The main drawback of finite loading relates to its schedules becoming obsolete from unanticipated materials delay

and inaccurate processing time estimates. It has to be updated frequently and the cost of updating a finite loading system is much more than sequencing by priority rutes in an infinite loading system. The main advantage is that accurate capacity and load estimates for the short term for the next few days are obtained.

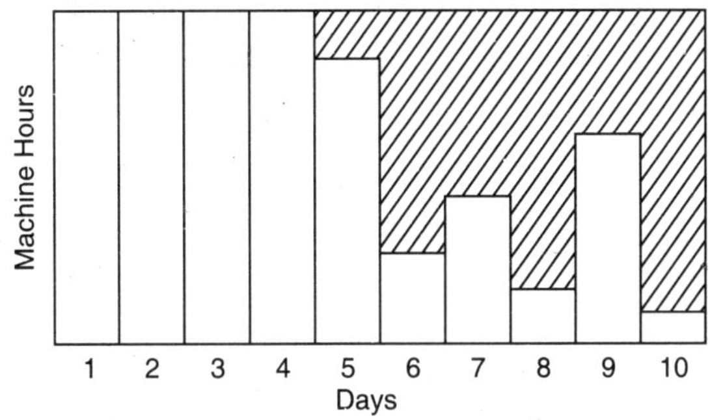

Fig. 8.3 Finite loading of work centre with a fixed capacity of 100 machine hours.

8.2.2 Assignment Model

Managers often face the problem of deciding optimal assignment of jobs to facilities which will maximize profits or minimize costs. Loading of facilities for optimum utilization involves assigning n jobs to n facilities so as to maximize profits or minimize costs or processing time. The assignment algorithm provides a simple procedure for optimum loading of facilities. It is based on the following assumptions:

(*i*) Number of jobs to be assigned must be equal to the number of facilities.

(*ii*) Each facility must be assigned one and only one job.

(*iii*) The criterion of determining a good or bad assignment must be specified, the criteria may be profits, operating cost, processing time etc.

(*iv*) The algorithm provides for assignment analysis of a job at one point of time and is not applicable when the jobs are arriving continuously.

The procedure adopted in an assignment algorithm consists of the following steps (Hungarian Method)

1. Subtract the smallest number in each row from all numbers in that row so that there is at least one zero in each row.
2. Subtract the smallest number in each column from all numbers in that column.
3. Check whether an optimal assignment can be made in the reduced matrix or not. Starting from first row/column, determine the row/column with only one zero and make an assignment (□) corresponding to that zero. Delete that row and column where the assignment has been made. Determine another row/column with single zero in the remaining matrix and make an assignments again, the procedure is repeated till all the assignments have been made. If we find that there is any row or column without assignment then we proceed to step 4.
4. Mark (√) the rows for which assignment has not been made.
5. Mark columns which have zeros in marked rows.

6. Mark rows which have assignments in marked columns.
7. Repeat steps 5 and 6 until no more marking is possible.
8. Draw lines through all unmarked rows and through all marked columns. This gives the minimum number of lines crossing all zeros. If procedure is correct, there will be as many lines as the number of assignments.
9. Examine the elements that do not have a line through them. Select the smallest of these elements and subtract it from all the elements that do not have a line through them. Add this smallest element to every element that lies at the intersection of two lines. Leave the remaining elements of the matrix unchanged.
10. Check if optimal assigment can be made in the current feasible solution or not. Repeat step 111.

Example 8.1: *Four different jobs can be done on four different machines. The set-up and take-down time costs are assumed to be prohibitively high for changeovers. The matrix below gives the cost in rupees of producing Job i on machine J.*

Machines

Jobs		M_1	M_2	M_3	M_4
	J_1	5	7	11	6
	J_2	8	5	9	6
	J_3	4	7	10	7
	J_4	10	4	8	3

How should the jobs be assigned to the various machiens so that the total cost is minimized.

Solution: Step I. Prepare a square matrix. In this case it is a square matrix.

Step II. Reduce the Matrix

	M_1	M_2	M_3	M_4
J_1	0	2	6	1
J_2	3	0	4	1
J_3	0	3	6	3
J_4	7	1	5	0

→

J_1	0	2	2	1
J_2	3	0	0	1
J_3	0	3	2	3
J_4	7	1	1	0

Step III. Optimal Assignment

	M_1	M_2	M_3	M_4
J_1	[0]	2	2	1
J_2	3	[0]	~~0~~	1
J_3	~~0~~	3	2	3
J_4	7	1	1	[0]

Step IV Mark (√) rows for which assignment has not been made.

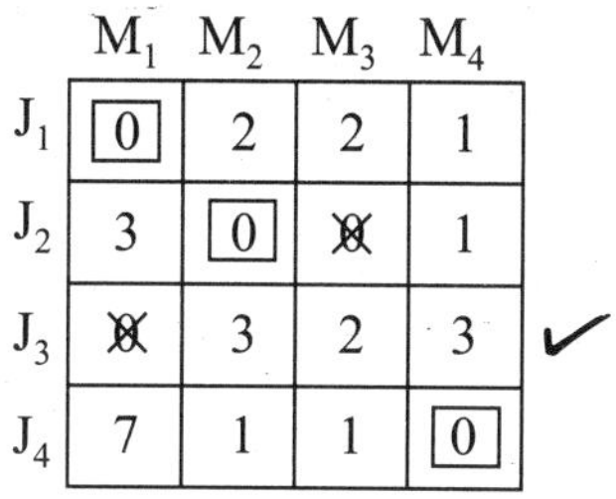

	M_1	M_2	M_3	M_4	
J_1	[0]	2	2	1	
J_2	3	[0]	⊗	1	
J_3	⊗	3	2	3	✓
J_4	7	1	1	[0]	

Step V. (*i*) Mark columns which have zeros in marked rows.

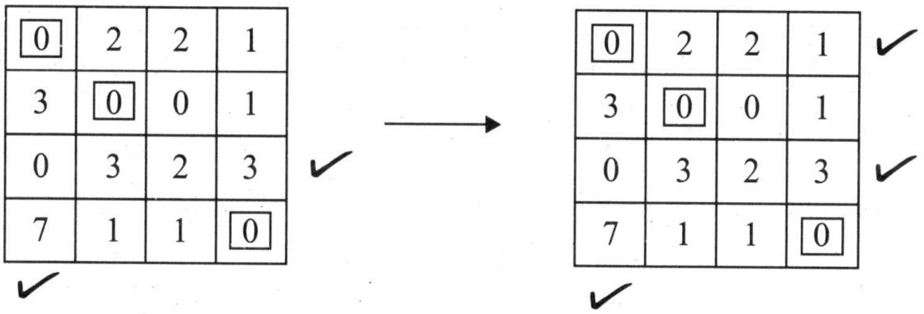

[0]	2	2	1	
3	[0]	0	1	
0	3	2	3	✓
7	1	1	[0]	
✓				

→

[0]	2	2	1	✓
3	[0]	0	1	
0	3	2	3	✓
7	1	1	[0]	
✓				

(*ii*) Mark rows which have assignments in marked column.

Step VI. (*i*) Mark lines covering unmarked rows and marked columns.

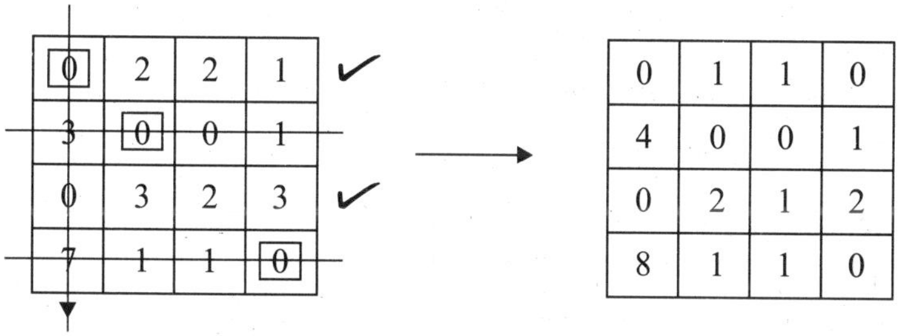

(Lines drawn through column 1 and through rows 2 and 4.)

[0]	2	2	1	✓
3	[0]	0	1	
0	3	2	3	✓
7	1	1	[0]	

→

0	1	1	0
4	0	0	1
0	2	1	2
8	1	1	0

(*ii*) Subtract lowest element from all uncovered elements and add lowest element to intersecting element.

Step VII. Give optimal assignment. Repeat steps IV, V, VI.

⊗	1	1	[0]
4	0	0	1
[0]	2	1	2
8	1	1	⊗

Step VIII.

(Lines drawn through columns M_1 and M_4 and through row J_2.)

	M_1	M_2	M_3	M_4	
J_1	⊗	1	1	[0]	✓
J_2	4	[0]	⊗	1	
J_3	[0]	2	1	2	✓
J_4	8	1	1	⊗	✓
	✓			✓	

→

[0]	⊗	⊗	⊗
5	[0]	⊗	2
⊗	1	[0]	2
8	⊗	⊗	[0]

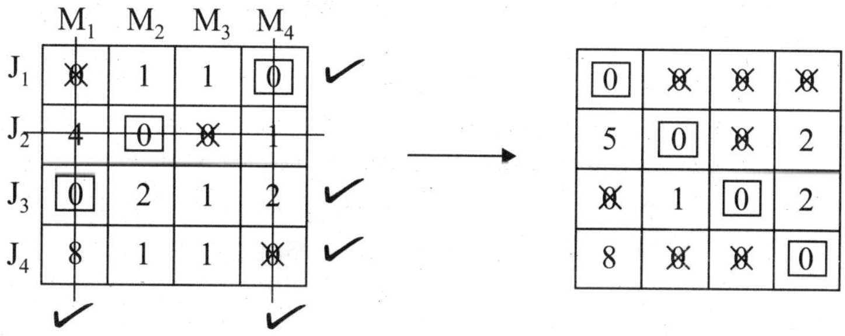

So optimum solution is $J_1 - M_1$, $J_2 - M_2$, $J_3 - M_3$, $J_4 - M_4$
and optimum cost = 5 + 5 + 10 + 3 = ₹ 23

Example 8.2: *A company has a team of four salesman and there are four districts where the company wants to start its business. After taking into account the capabilities of salesman and the nature of districts, the company estimates that the profit per day in rupees for each salesman in each district is as below.*

Find the assignment of salesman to various districts which will yield maximum profit.

Salesman \ *District*	*1*	*2*	*3*	*4*
A	*16*	*10*	*14*	*11*
B	*14*	*11*	*15*	*15*
C	*15*	*15*	*13*	*12*
D	*13*	*12*	*14*	*15*

Solution: First convert maximization problem into minimization problem by subtracting all the elements from the highest element and then apply Hungarian Method as mentioned earlier.

	1	2	3	4
A	0	6	2	5
B	2	5	1	1
C	1	1	3	4
D	3	4	2	1

→

	1	2	3	4
A	0	6	2	5
B	1	4	0	0
C	0	0	2	3
D	2	3	1	0

→

	1	2	3	4
A	[0]	6	2	5
B	1	4	[0]	~~0~~
C	~~0~~	[0]	2	3
D	2	3	1	[0]

So optimal solution is A – 1, B – 3, C – 2, D – 4
16 + 15 + 15 + 15 = ₹ 61.

8.3 SCHEDULING

Scheduling can be defined as "the process of filling a production order into the available time of the machine or of the work centre, and within a total time which is within the promised delivery schedules."

Objectives:

(*i*) To ensure maximum utilization of the plant at minimum cost.

(*ii*) To ensure that requirements of manpower is optimum and is evenly distributed, there being no peaks and valleys.

(*iii*) To keep up-to-date information on the available workforce.

(iv) To possess up-to-date information regarding the availability of materials, expected delivery dates, materials rejection, purchase order cancelled, etc.

(*v*) To possess information on each machine regarding its break-down, availability of spares, average frequency of breakdowns, servicing and overhauling schedules, replacement schedules etc.

(*vi*) To possess complete information on performance standards and their revisions, method improvement, changes in materials and machines etc.

(*vii*) To obtain a quick feedback from shops regarding delays and interruptions which may hold up production activity.

8.3.1 Forms of Scheduling

There are various forms of scheduling as given below:

1. **Master Scheduling:** It refers to overall scheduling of weekly production of all end products to be manufactured by a plant to meet customer orders and demand forecasts. It is simply a weekly or monthly breakdown of the production requirements of each production for a definite period of time (a quarter, six months, or a year). As new orders or requirements are received, they are scheduled on the master schedule with due regard to available plant capacity. The flexibility of master scheduling is a desirable feature because firm master schedules remain unrealized ideals for most of the companies, whether manufacturing to stock or to customer's orders. When the company manufacturers to stock, a change in master schedule occurs every time sales do not follow the forecast closely. When it manufactures to customer orders, the master schedule has to follow a customer's dictates.

2. **Operation Scheduling:** It consists of assigning existing orders and or imminent forecast to specific resources (facilities, employees and equipment) and then sequencing the jobs at each processing centre to achieve optimum utilization of existing capacity.

3. **Detailed Scheduling:** It refers to assigning of the time required to do each detailed operation of a given job with a given machine or process.

8.3.2 Approaches of Scheduling

There are two approaches of scheduling that is forward scheduling and backward scheduling.

1. **Forward Scheduling:** Forward scheduling is commonly used in job shops. It refers to determining the start and finish time for waiting jobs by assigning them to the earliest available time slots at the work centre. As the job and its components start as early as possible, they are likely to be completed before they are due at subsequent work centres in the routing. Consequently forward scheduling procedure accumulates in process inventories that remain throughout the facility until they are needed at subsequent stations. These excessive inventories are a drawback of forward scheduling. The advantages of forward scheduling include its simplicity and short lead times in getting the jobs completed.

2. **Backward Scheduling:** Backward scheduling is often used in assembly type industries and in job shops that commit in advance, to specific delivery dates. It refers to determining the start and finish times for waiting jobs by assigning them to the latest available time slot that will enable each job to be completed just when it is due, but not before. The job start time is determined by setting back from the finish date. By assigning jobs as late as possible, backward scheduling minimizes inventores since a job is not completed until it must go directly to the next work centre on its routing. It requires that bill of materials aid lead time estimates must be accurately maintained for all facilities or else the system break down, due dates are violated and delivery service to customers deteriorates.

8.4 SCHEDULING SYSTEMS

Scheduling systems may be classified as follows:

1. **Job Shop Scheduling System:** This is used for scheduling when jobs are produced one-by-one and are of different types, that is for job production.
2. **Batch Scheduling System:** When jobs are produced to order in batches, this is used.
3. **Mass Scheduling System:** When large number of items of similar type are produced that is in mass production, this is used.

8.4.1 Job Shop Scheduling System

In job shop scheduling, we come across varities of jobs to be processed on different types of machines. Separate records are to be maintained for each order. Only after receiving the order one has to plan for production of the job. The routing is to be specified only after taking the order. Scheduling is done to see that the available resources are used optimally. The following are the factors on which job shop scheduling depends:

(*i*) Number of jobs to be scheduled.

(*ii*) The number of machines in the shop.

(*iii*) Flow patterns of the jobs through the shop—When all jobs follow the same path from one machine to the next, it is referred to as *flow shop* and when there is no similar pattern of movement from one machine to next, it is called a *job shop*.

(*iv*) Job arrival patterns—The pattern is termed static when a number of jobs with known processing times require processing on machines and the new jobs arrivals do not disturb the processing of these jobs. Such a situation exists when the jobs are scheduled after a fixed time interval, say once a week. It means that jobs are not scheduled unless all the previous week's incoming orders are received. The pattern is termed as dynamic, when jobs are scheduled as they arrive and the overall schedule is updated to reflect their effect on production facilities.

(*v*) Priority rules for allocating jobs to machines: A priority rule is a rule for selecting which job is to be started first on some machine or work centre. Examples of priority rules are : First Come First Served (FCFS), Shortest Processing Time (SPT), Earliest Due Date (EDD), etc.

(*vi*) Schedule evaluation criteria—A satisfactory schedule may be evaluated in terms of the satisfactory completion of jobs, utilization of productive facilities and meeting overall organization objectives.

Techniques of Job Shop Scheduling: The various techniques of job shop scheduling are discussed below:

(*a*) **Order Scheduling:** In this method, for each and every operation assigned to a facility, the-set up time, starting time, completion time; and in many cases, even the material movement times are specified. In order scheduling by Gantt chart, the horizontal axis represents the time interval in hours, work days or work weeks.

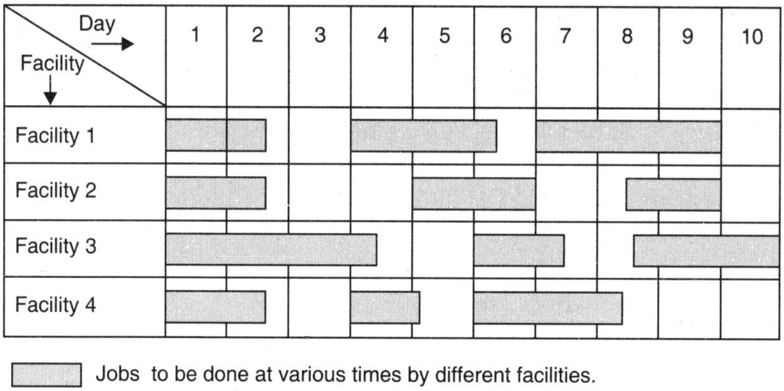

Fig. 8.4 Order scheduling by Gantt chart.

The selection of time interval depends upon the individual situation. Figure 8.4 shows an order schedule Gantt chart. The information required for each job to be scheduled into production by means of an order schedule chart is listed below.

(*i*) The number of items to be produced.

(*ii*) The number of machine assigned to do the work.

(*iii*) The set-up time necessary for each operation.

(*iv*) The total production time necessary to do each operation.

(*v*) The required date of completion for the job.

(*vi*) The exact status of every job already scheduled.

The scheduling of a new job can begin, once the basic information is known. The scheduler starts with the last operation placed on the schedule board at the required completion date and then works backwards to the first operation. The time available in the beginning, if any, is used for rush jobs. However, if there are no rush jobs, the starting times of the jobs can be advanced as required. Order scheduling is basically a trial process in which the various operations are arranged and rearranged until an acceptable overall plan has been devised.

Advantages

- Each and every phase of operation is planned in detail.
- The status of every job in a shop can be predicted with a high accuracy.

Disadvantages

- It is a costly method of planning the activity.
- It is difficult to monitor the progress when operation times are short.
- It is difficult to maintain when there are many active orders.
- It is difficult to maintain when there are many operations on each order.
- Accurate time standards must be available.
- It is difficult to maintain when frequent changes are necessary.
- Communications within the organization need to be highly developed.

(*b*) **Index Method of Scheduling:** The standard method of scheduling jobs on different facilities is on a first come first served basis. In this method, jobs are scheduled as they arrive in the scheduling department to the best available machine at that time.

This method is suitable when there is a small backlog of jobs. However, in case it is possible to accumulate a small group of jobs, the jobs can be assigned to the facilities in a better way by using index method. It involves the following steps:

(*i*) The jobs to be scheduled on a number of facilities are accumulated in the form of a backlog. The larger the number of jobs accumulated, the larger is the saving that would result by using index method of scheduling.

(*ii*) The jobs are arranged in a table along with their processing times on the facilities on which they need to be scheduled. The bottom of the columns provide the available facility times.

(*iii*) The jobs are assigned to the best facility regardless of the machine time availability.

(*iv*) An index number is obtained by dividing the processing time on a non-optimal facility by the processing time taken on the best facility. This index is used as a guide for rearranging the assignments in order to obtain a best possible schedule.

(*v*) The samllest index number is determined for an alternative facility. After removing the job from the optimal facility it is reassigned to the alternative facility with the smallest index number.

(*vi*) The procedure of reassigning is repeated till an optimal schedule is obtained.

The procedure of index method of scheduling is explained in the following example:

Example 8.3: *The processing times of 10 jobs on 4 machines A, B, C and D are shown in Table 8.2. The processing times on machine A are less as the machine is new and the processing times on machine D are more as the machine is quite old. The available hours on machines A, B, C and D are 65 each. Schedule the jobs using the standard method. Use index method to improve the schedule.*

Table 8.2 Processing Time in Hours

Job Number	*Machine A*	*Machine B*	*Machine C*	*Machine D*
1	10	15	14	12
2	18	20	22	27
3	17	21	25	28
4	16	17	24	25
5	12	20	17	Not possible
6	16	22	19	28
7	12	Not Possible	18	22
8	15	18	16	20
9	25	30	27	35
10	18	25	29	32
Total Hrs Available	65	65	65	65 Total 260

Standard Method of Scheduling: The first 4 jobs are assigned to machine A which will require 10 + 18 + 17 + 16 = 61 hours. Job 5 cannot be assigned to machine

A as the total hours available on machine A are only 65 hours. Jobs 5, 6 and 7 are assigned to machine requiring 17 + 19 + 18 = 54 hours. Machine *C* has been selected for jobs 5, 6 and 7 as it has lower processing times as compared to machine *B* and *D*. Job 8 cannot be assigned to machine *C* because the total hours available on machine *C* are also 65 and assigning job 8 on machine *C* would require 70 hours. Jobs .8 and 9 are assigned on machine B requiring 48 hours. These jobs are not assigned on machine *D* as the processing time on machine *B* is less. The last job 10 is assigned to machine *D* requiring 32 hours. Thus, the total hours required as a result of standard method of scheduling can be calculated as

= 61 h on machine A + 48 h on machine B + 54 h on machine C
+ 32 h on machine D = 195 hours.

	Processing Time in Hours			
Job Number	***Machine A***	***Machine B***	***Machine C***	***Machine D***
1	10	15 (1.50)	14 (1.40)	12 (1.20)
2	18	20 (1.11)	22 (1.22)	27 (1.50)
3	17	21 (1.25)	25 (1.47)	28 (1.65)
4	16	17 (1.06)	24 (1.50)	25 (1.56)
5	12	20 (1.67)	17 (1.42)	Not Possible
6	16	22 (1.38)	19 (1.18)	28 (1.75)
7	12	Not Possible	18 (1.50)	22 (1.83)
8	15	18 (1.20)	16 (1.06)	20 (1.33)
9	25	30 (1.20)	27 (1.08)	35 (1.40)
10	18	25 (1.39)	29 (1.62)	32 (1.78)

The index numbers are shown in brackets. The sample calculation of index number is as follows:

The index number for machine B corresponding to job 1

$$= \frac{\text{Processing Time on Machine B}}{\text{Processing Time on Optimal machine, } i.e., \text{ A}} = \frac{15}{10} = 1.5$$

Similarly, the other index numbers can be calculated.

	Machine A	***Machine B***	***Machine C***	***Machine D***
Starting Assignment-all jobs on machine A	159	0	0	0
First change shifting job 8 on machine C	– 15		+ 16	
	144	0	16	0
Second change shifting job 4 on machine B	– 16	+ 17		
	128	17	16	0
Third change shifting job 9 on machine C	– 25		+ 27	
	103	17	43	0

Fourth change shifting job 2 on machine B	– 18	+ 20		
	85	37	43	0
Fifth change shifting job 6 machine C	– 16	0	+ 19	0
	69	37	62	0
Sixth change shifting job 1 on machine D	– 10	0	0	+ 12
	59	37	62	12

In order to use the index method, as an initial solution first assign all the jobs to the best machine, *i.e.,* machine *A*. It would require a total of 159 hours of processing time on machine A and 0 processing time on machines B, C and D. However, this is not feasible solution as the maximum available time on machine *A* is 65 hours.

Modify the initial solution by determining the minimum index number in the table, which happens to be 1.06 for machine C corresponding to job 8 and for machine *B* corresponding to job 4. The processing time for machine C for job 8 is 16 hours which is less than the processing time of job 4 on machine *B* which is 17 hours. Therefore, in the first shift of assignment, job 8 is shifted to machine *C* from machine A, resulting in a reduction of 15 hours of processing time on machine A (job 8 requires 15 hours of processing on machine A). Now removing job 8 from the assignment list, the next minimum processing time of 17 hours occurs on machine B corresponding to job 4 for an index number of 1.06. In the second change, shift job 4 from machine *A* to machine B. Proceeding in this way, the changes can be made till the constraint of available hours on machine *A* and other machines is satisfied. This results in the following schedule:

The total time required by making use of index method of scheduling works out to be .

59 hours on machine A + 37 hours on machine *B*

+ 62 hours on machine C + 12 hours on machine D = 170 hours

The index method of scheduling thus results in a saving of 25 hours (195 – 170) over the standard method of scheduling.

8.4.1.1 Sequencing

The selection of an appropriate order for a series of jobs to be done on a finite number of service facilities is called sequencing. Priority sequencing rule refers to a systematic procedure for assigning priorities to waiting jobs, thereby determining the sequence in which jobs will be processed. Thevarious factors which influence the selection of a particular sequencing rule relate to (*i*) size of inventories, (*it*) set-up costs, (*iii*) idle time, (*iv*) number of per cent of jobs that are late, (*v*) average time that jobs are late by, (*vi*) standard deviation of the time that jobs are late, (*vii*) average number of jobs waiting in the queue, (*viii*) average time to complete a job, and (*ix*) standard deviation of time to complete a job.

Three of these criteria namely set-up costs, inventory costs and idle time are primarily concerned with internal facility efficiency. Minimizing these results in better use of limited resources and improved profitability for the organization. Three of the criteria namely per cent of jobs that are late, average time that jobs are late and variance of time that jobs are late are more customer oriented and minimizing them results in better service to the customers. Finally, three of the criteria namely number of jobs waiting, average time to complete a job and variance of time to complete a job, reflect both customer service as well as internal efficiency. To find a sequencing rule that maximizes all of these criteria is quite difficult.

The following assumptions are usually made when dealing with sequencing problems:

- Only one operation is carried out on a machine at a particular time.
- Each operation, once started, must be performed till completion.
- Each operation must be completed before its succeeding operation can start.
- There is only one machine of each type available in the shop.
- Job is processed as soon as possible, but only in the order specified.
- Processing times are independent of order of performing the operations.
- The time required to transfer jobs between machines is negligible.
- All jobs are completely known and are ready to start processing before the period under consideration starts.

The important priority sequencing rules followed in productive/service systems are given below:

First Come First Served (FCFS): It is a priority rule that gives top priority to the waiting job that arrived earliest in the productive system. It is quite widely used in service industries such as banks, hospitals, railway booking offices, etc. and provides for general agreement and satisfaction amongst the customers.

Earliest Due Date (EDD): In this rule, top priority is assigned to the waiting job whose due date is earliest. It ignores when the job arrives and the time takes to process.

Shortest Processing Time (SPT): According to this priority rule, top priority is given, to the waiting job whose operation time at a work centre is shortest. The jobs, due dates and order of arrival are immaterial.

Least Slack Rule (LS): It gives top priority to the waiting job whose slack time is the least. Slack time is the difference between the length of time remaining until the job, is due and the length of its operation time.

Average job flow time: It is defined as the total time spent by a job in the system.

Average Number of Jobs in the System: It is defined as the average number of jobs flowing in the system (waiting or being processed) from the beginning of sequence through the time when the last job is finished.

Average Job Lateness: Job lateness is defined as a difference between the actual completion time of the job and its due date. Average job lateness is sum of lateness of all jobs divided by the number of jobs in the system. This is also referred to as average job tardiness.

Average Earliness of Job: If a job is completed before its due date, the lateness value is negative and the magnitude is referred to as earliness of job. Mean earliness of the job is the sum of earliness of jobs divided by the number of jobs in the system.

Number of Tardy Jobs: It is number of jobs which are completed after the due date.

Example 8.4: *The following table shows the processing times for five jobs, and the due date for the jobs in a machine shop. The jobs are to be processed based on FCFS, SPT, EDD, and LS priority rules. Determine (i) Total completion time, (ii) Average Flow Time, (iii) Average number of jobs in the Machine shop and (iv) Average Job Lateness, (v) Average Job Earliness and (vi) Number of Tardy Jobs based on each of these priority sequencing rules and make a comparison.*

Jobs to be Processed	*Processing Time (in days)*	*Due Date (in days with reference to start of processing)*
A	*4*	*6*
B	*17*	*20*
C	*14*	*18*
D	*9*	*12*
E	*11*	*12*

Solution: (*a*) FCFS Priority Rule: As the jobs are to be sequenced based on *FCFS* priority rule, the following job sequence results:

Waiting Jobs in FCFS Sequence	*Processing Time (in days)*	*Flow Time (in days)*	*Due Date (in days with reference to start of processing)*
A	4	4	6
B	17	21	20
C	14	35	18
D	9	44	12
E	11	55	12
	Total 55 days		

The total completion time of all jobs = 4 +17 + 14 + 9 + 11 = 55 days

The average flow time = (4 + 21. + 35 + 44 + 55)/(5) = 31.8 days

Average Number of Jobs in the Shop: The average number of jobs in the shop (waiting or being processed) can be worked out as follows:

For the first four days, 5 jobs are in the shop; for the next seventeen days, 4 jobs are in shop, for days twenty-two to thirty-five, 3 jobs are in; days thirty-six to forty-four, 2 jobs are in the shop and days forty-five to fifty-five only 1 job is in the system. Therefore average no. of jobs in the system:

$$= [5 \times 4 + 4 \times 17 + 3 \times 14 + 2 \times 9 + 1 \times 11]/55 = 2.89 \text{ Jobs.}$$

Average Job Lateness can be worked out as follows:

The lateness of each job is the difference of its flow time and its due date. Flow time for job A is 4 days and its due date is 6 days. The difference is 4 – 6 = – 2 which means job A has finished two days earlier than required, *i.e.,* no lateness or zero lateness. Flow time for job B is 21 days and its due date is 20 days, the difference is 21 – 20 = 1 day, *i.e.,* the job is late by one day. Similarly, lateness for jobs C, D and E can be calculated as 17, 32 and 43 days, respectively. The average lateness is therefore given as

$$= (0 + 1 + 17 + 32 + 43)/(5) = 18.6 \text{ days}$$

Average Job Earliness: Only job *A* finishes earlier than the due date by two days and all other jobs finish late. The average job earliness can be calculated by considering job earliness for jobs B, C, D and E as zero, which is given as follows:

Average job earliness = (2 + 0 + 0 + 0 + 0)/5 = 0.4 day

Number of tardy jobs = 4

(*b*) ***Shortest Processing Time (SPT) Sequencing Rule:*** The sequencing of jobs and flow time based on SPT sequencing rule is shown below.

Waiting Jobs in SPT Sequence	*Processing Time (in days)*	*Flow Time (in days)*	*Due Date (in days with reference to start of processing)*
A	4	4	6
D	9	13	12
E	11	24	12
C	14	38	18
B	17	55	20
Total	55 days		

The total processing time remains same, *i.e.*, 55 days

The average flow time is given as

$$= (4 + 13 + 24 + 38 + 55)/(5) = 26.8 \text{ days}$$

The average number of jobs in the shop are :

$$= (5 \times 4 + 4 \times 9 + 3 \times 11 + 2 \times 14 + 1 \times 17)/(55) = 2.44 \text{ jobs}$$

Average Job Lateness: The lateness for jobs in this sequence is 0, 1, 12, 20 and 35 days for jobs A, B, C, D and E, respectively. Therefore average job lateness is

$$= (0 + 1 + 12 + 20 + 35)/(5) = 13.6 \text{ days}$$

Average Job Earliness: The earliness of jobs A, B, C, D and E can be worked out as 2, 0, 0, 0, and 0 respectively. Therefore average job earliness is given as

$$= (2 + 0 + 0 + 0 + 0)/5 = 0.4 \text{ days}$$

Number of tardy jobs = 4

(*c*) ***EDD Sequencing Rule:*** The sequencing of jobs based on this rule are as follows:

Waiting Jobs in EDD Sequence	*Processing Time (in days)*	*Flow Time (in days)*	*Due Date (in days with reference to start of processing)*
A	4	4	6
D	9	13	12
E	11	24	12
C	14	38	18
B	17	55	20
Total	55 days		

As the *EDD* sequencing rule happens to result in the same sequence as given by *SPT* sequencing rule, all the parameters for *EDD* sequencing rule will remain same as calculated by using SPT sequencing rule.

(*d*) ***Least Slack Sequencing Rule (LS):*** The slack time corresponding to jobs A, B, C, D and E can be worked as 2, 3, 4, 3, and 1 day, respectively. The sequencing of jobs based on least slack and the flow time for the job is as given below:

Waiting Jobs in LS Sequence	*Processing Time (in days)*	*Flow Time (in days)*	*Due Date (in days with reference to start of processing)*	*Slack time (in days)*
E	11	11	12	1
A	4	15	6	2
B	17	32	20	3
D	9	41	12	3
C	14	55	18	4
	Total 55 days	154		

The average flow time of jobs = (11 + 15 + 32 + 41 + 55)/5 = 30.8 days

Average number of jobs in the system

$$= (5 \times 11 + 4 \times 4 + 3 \times 17 + 2 \times 9 + 1 \times 14)/55 = 2.8 \text{ jobs.}$$

Average Job Earliness: The job earliness for jobs E, A, B, D and C can be worked as 1, 0, 0, 0 and 0, respectively. Therefore average job earliness is given as

$$= (1 + 0 + 0 + 0 + 0)/5 = 0.2 \text{ days.}$$

Average Job Lateness: The Job lateness for jobs E, A, B, D and C can be worked as 0, 9, 12, 29, and 37 days. Therefore, Average Job Lateness is given as

$$= (0 + 9 + 12 + 29 + 37)/5 = 17.6 \text{ days.}$$

Number of Tardy Jobs = 4.

Parameter	*FCFS*	*SPT*	*EDD*	*LS*
Total completion time (days)	55.0	55.0	55.0	55.0
Average flow time (days)	31.8	26.8	26.8	30.8
Average no. of jobs	2.89	2.44	2.44	2.80
Average job lateness (days)	18.6	13.6	13.6	17.6
No of tardy jobs	4	4	4	4
Average job earliness	0.4	0.4	0.4	0.2

Job Shop Sequencing Problems

Job shop sequencing problems can be of various types. Some of the job shop sequencing problems are described here:

Sequencing n Jobs on one Machine: Sequencing problems can arise even when only one service facility is involved. For example, number of cars to be serviced at a work station, number of patients to be seen by a doctor, number of job requests from different terminal users to be executed by a computer, number of different jobs to be machined on a lathe, and so on. The job can be sequenced by selecting an appropriate sequencing rule such as FCFS, SPT, EDD and LS, etc. as described earlier.

Sequencing n Jobs through Two Machines: The sequencing algorithm for optimum sequencing of *n* jobs through two machines so that the total make span time of all the jobs is minimum was given by Johnson and is known as Johnson's Algorithm. It is based on the following assumptions:

(*i*) There are only two machines and the processing of all the jobs is to be done on both the machines in the same order, *i.e.,* first on machine 1 and then on machine 2.

(*ii*) All jobs arrive at the same time and have no priority rule for job completion.

The method for Johnson's algorithm involves following steps:

(*i*) List operations time for each job on the two machines.

(*ii*) Select the shortest operation time in the list.

(*iii*) If the shortest time happens to be on machine 1, sequence the corresponding job as early as possible. If it is for machine 2, sequence the corresponding job as late as possible. In case of tie in shortest operation time, it can be broken arbitrarily.

(*iv*) Eliminate both the operation time and the job which has already been sequenced in step (*iii*).

(*v*) Repeat steps (*ii*) and (*iii*) to complete sequencing of the remaining jobs.

Example 8.5: *Six jobs are to be sequenced which require processing on two machines. The processing time in minutes for each job on these machines is given below. All the jobs are to be processed first on machine 1 and then machine 2. Determine the optimum sequence for processing the jobs in the machine shop so that the total make span time of all the jobs is minimum using Johnson's algorithm.*

Jobs	*1*	*2*	*3*	*4*	*5*	*6*
Processing time on Machine-1	*30*	*30*	*60*	*20*	*35*	*45*
Processing time on Machine-2	*45*	*15*	*40*	*25*	*30*	*70*

Solution: Step 1. Determine the minimum processing time in the table. On examining the table, the minimum processing time is 15 minutes on machine 2 for job 2. As the minimum time happens to be on machine 2, it is sequenced in the last as follows:

					2

Now minimum processing time is 20 min on machine 1 for job 4.

4					2

Proceeding in this manner the next minimum processing time is 30 minutes for job 5 on machine 2 and job 1 for machine 1, resulting in the following sequence

4	1			5	2

The next processing time is 40 minutes for job 3 on machine 2 resulting in the following sequence:

4	1		3	5	2

Lastly the job 6 can be sequenced in the remaining slot resulting in the following optimal sequence of jobs:

4	1	6	3	5	2

The minimum make span time of all the jobs utilizing this optimal sequence, *i.e.,* 4-1-6-3-5-2 can be determined by drawing a Gantt chart as shown in Fig. 8.5. As shown in Fig. 8.5 the minimum make span time for the six jobs on the two machines is 250 minutes.

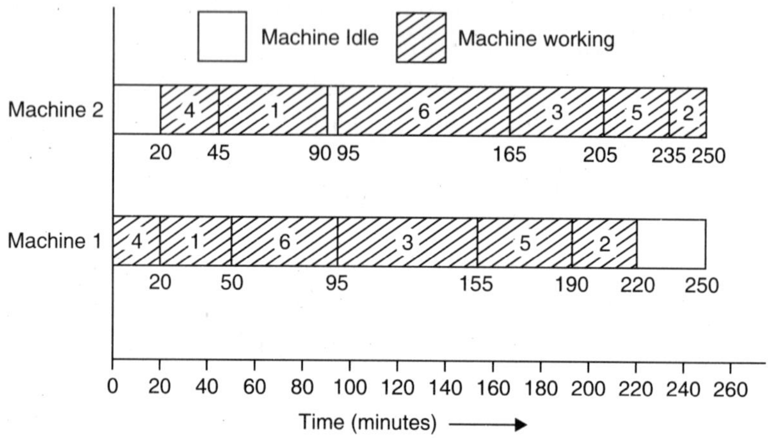

Fig. 8.5 Gantt chart for determining minimum make span time.

The Fig. also depicts the idle time for machine 1 (30 Minutes) after finishing the processing of last job 2 and the idle time for machine 2 (25 minutes; 20 minute before staring processing of job 4 and 5 minutes before processing of job 1 after finishing job 4).

The percentage utilization of machine 1 and machine 2 can be worked out as follows:

$$\text{Machine 1} = \frac{250-30}{250} = 88\%, \text{ Machine 2} = \frac{250-25}{250} = 90\%$$

Jackson's Modified Algorithm: Jackson modified algorithm provide a minimum make span sequence of *n* jobs in a two machine job shop in which jobs may have different processing sequences. The jobs are classified into 4 sets which are defined as follows:

[A] = The sets of jobs which are processed only on machine 1

[B] = The set of jobs that are processed only on machine 2

[AB] = The set of jobs to be processed on machine 1 followed by machine 2

[BA] = The set of jobs to be processed on machine 2 followed by machine 1

After classifying the jobs into the above sets, the jobs in set [AB] are sequenced by Johnson's algorithm. Then the jobs in set [BA] are sequenced by Johnson's algorithm and finally any arbitrary sequence is selected for the jobs in set [A] and in set [B]. The sets of jobs are then combined in the following way, without changing the order within each set:

Machine 1: Jobs in [AB] before jobs in [A] before jobs in [BA]
Machine 2: Jobs in [BA] before jobs in [B] before jobs in [AB]

Example 8.6: *A set of ten jobs is to be processed through a two machine job shop. The technological order and processing times of the jobs are shown below. Use Jackson's extention of Johnson's algorithm to find a minimum make span schedule. Construct a Gantt chart of the schedule and determine the make span and machine idle time.*

	Machine Number for		*Processing Times*	
Job	***Operation 1***	***Operation 2***	***Operation 1***	***Operation 2***
1	1	2	4	6
2	1		3	
3	1		4	
4	1	2	5	2
5	2	1	1	2
6	2		1	
7	2	1	7	8
8	2		3	
9	2	1	6	7
10	1	2	2	4

Solution: Set of jobs only to be processed on machine 1 [2, 3]
Set of jobs only to be processed on macnine 2 [6, 8]
Set of jobs to be processed on machine 1 followed by machine 2 [2, 4, 10]
Set of jobs to be processed on machine 2 followed by machine 1 [5, 7, 9]

Considering jobs 1, 4 and 10

Job	*1*	*4*	*10*
Processing time on Machine 1	4	5	2
Processing time on Machine 2	6	2	4

The optimal sequence for these jobs using Johnson's algorithm is given as

10	1	4

Considering jobs 5, 7 and 9

Job	*5*	*7*	*9*
Processing Time on Machine 1	2	8	7
Processing Time on Machine 2	1	7	6

The optimal sequence for these jobs using Johnson's algorithm is given as

7	9	5

Therefore, using Jackson's extension of Johnson's algorithm the sequence of jobs on machine 1 and machine 2 would be as follows:

Sequence of jobs on machine 1 10—1—4—2—3—7—9—5

Sequence of jobs on machine 2 7—9—5—6—8—10—1—4

The minimum make span time for these 10 jobs is shown in Fig. 8.6. As shown in the figure minimum make span time is 35.

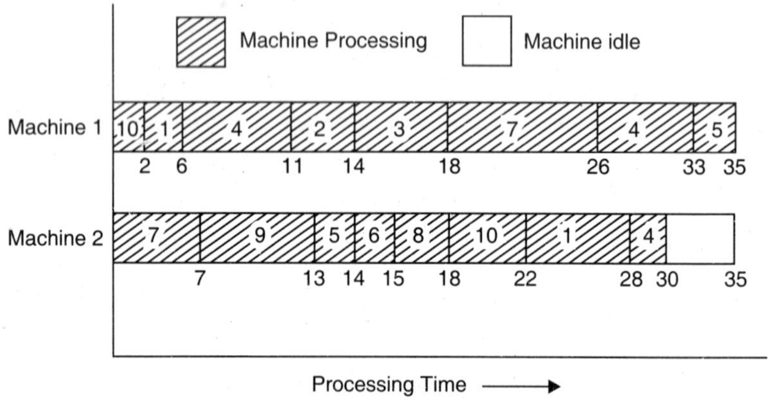

Fig. 8.6 Processing of 10 jobs on 2 machines using Jackson's extension of Johnson's algorithm.

The idle time for machine 1 is nil, and the idle time for machine 2 is 5.

The percentage utilization of machine 1 is 100% and the percentage utilization of machine 2 is (30/35) × 100 = 86%

Sequencing of n Jobs on 3 Machines: This sequencing problem relates to scheduling *n* jobs in 3 machines job shop. Johnson's algorithm for the 2 machines case may be extended to the 3 machines case if any one of the following conditions is satisfied:

Minimum processing time for a job on machine 1 is greater than or equal to the maximum processing time for a iob on machine 2 or minimum processing time for a job on machine 3 is greater than or equal to maximum processing time for a job on machine 2.

The assumptions made are as follows:

(*i*) Each job is processed on all the three machines in the same order, *i.e.,* first on machine 1, then on machine 2 and finally on *machine* 3.

(*ii*) The passing of jobs is not permitted.

(*iii*) The actual or expected processing times of jobs on these machines are known.

The procedure used is as follows:

(*i*) Define two dummy machines 1' and 2' such that the processing times of various jobs on these machines can be calculated as follows:

Processing time of a job on Machine 1' = Processing time on machine 1 + Processing time on machine 2

Processing time of a job on Machine 2' = Processing time on machine 2 + Processing time on machine 3 (*ii*) Now apply Johnson's algorithm on these two dummy machines 1′ and 2′ to determine the optimal sequence of jobs.

Exampler 8.7: *The processing times of 4 jobs on three machines is given in the following table:*

Job	*Machine 1*	*Machine 2*	*Machine 3*
1	*8*	*2*	*4*
2	*5*	*4*	*5*
3	*6*	*1*	*3*
4	*7*	*3*	*2*

Determine the optimum sequence for processing these jobs on these 3 machines so that the total make span time is minimum.

Solution: As the minimum processing time for job 2 on machine 1 is greater than the maximum processing time for job 2 on machine 2, the Johnson's modified algorithm can be used.

Combining the processing times of machine 1 and 2 and machine 2 and 3 to form dummy machines 1′ and 2′, the following matrix of processing times results:

Job	*Machine 1′*	*Machine 2′*
1	10	6
2	9	9
3	7	4
4	10	5

Using Johnson's algorithm for 2 machines shop, the minimum processing time of four occurs on machine 2′, therefore job 3 is sequenced at the last. The next mirimum time is 5 for job 4 on machine 2′, therefore job 4 is also sequenced towards the end before job 3. The next minimum processing time is 6 on machine 2′. Therefore job 1 is sequenced towards the end before job 4. The final optimal sequence is

2	1	4	3

Sequencing 2 Jobs through m Machines: The sequencing of 2 jobs through *m* machines is carried out by using a graphical procedure which involves following steps:

(*i*) Two axes are drawn at right angles to each other representing job 1 and job 2. Processing times on job 1 are marked along horizontal axis and processing times on job 2 are marked along vertical axis.

(*ii*) The machine times for the two jobs on corresponding axes are laid out in the given technological order.

(*iii*) The processing times common to job 1 and job 2 are represented by a shaded region. It represents an unfeasible region as simultaneous processing of both jobs on the same machine is not possible.

(*iv*) The processing of both jobs is represented by a continuous path which consists of horizontal, vertical and 45 degree diagonal segments. The path starts at the lower left corner and stops at upper right corner while avoiding the shaded areas in the graph. Thus the path is not allowed to pass through shaded areas which correspond to operating both jobs concurrently on the same machine. The vertical movement along the path indicates idle time for job 1, while the horizontal movement along the path indicates idle time for job 2. The diagonal segment implies that both jobs are being processed on different machines at the same time. A feasible path that maximizes the diagonal movement minimizes the total processing time.

Example 8.8: *The processing times of 2 jobs 1 and 2 on five machines A, B, C, D and E along with the sequence is given in the following table. Determine the optimum sequence in which the jobs should be processed so that the total make span time is minimum.*

Job 1	***Sequence***	*A*	*B*	*C*	*D*	*E*
	Time hours	*1*	*2*	*3*	*5*	*1*
Job 2	***Sequence***	*C*	*A*	*D*	*E*	*B*
	Time hours	*3*	*4*	*2*	*1*	*5*

Solution: The processing times of jobs 1 and 2 are shown graphically in Fig. 8.7 on two axis. The machine times for the two jobs in the given sequence are also shown in the figure. The shaded areas A, B, C, D and E represent the areas which are unfeasible as they represent simultaneous processing of jobs 1 and 2 on these machines. The optimal processing path is shown in the figure. It also depicts the idle time for job 1. There is no idle time for job 2. The optimal processing sequence is as follows:

Job 1 precedes job 2 on machine A, job 1 precedes job 2 on machine B, job 2 precedes job 1 on machine C, job 1 precedes job 2 on machine D and job 2 precedes job 1 on machine A.

The minimum processing time for jobs 1 and 2 can be calculated as total processing time for job 2 + idle time for job 2, which is 15 + 0 = 15 hours or minimum processing time for job 1 + idle processing time for job 1, which is = 12 + 3 = 15 hours.

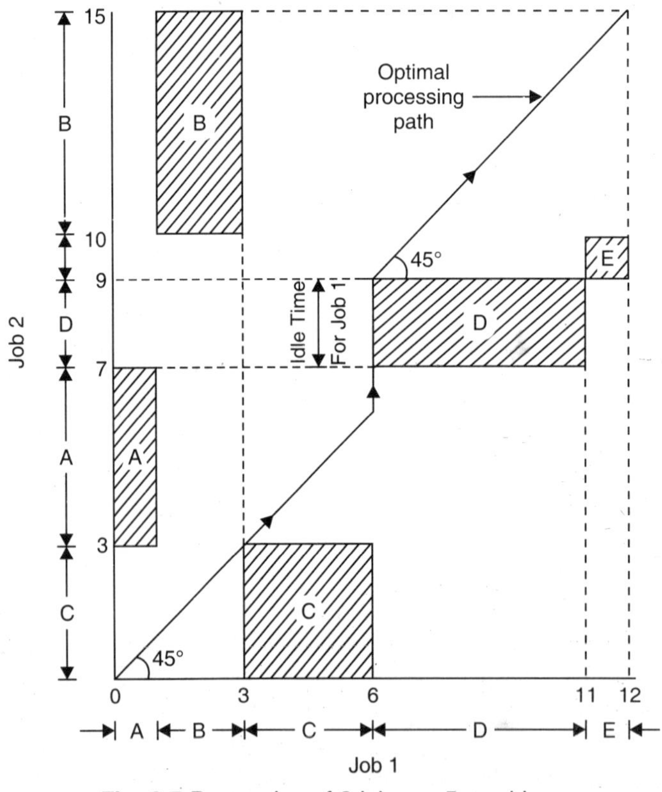

Fig. 8.7 Processing of 2 jobs on 5 machines.

8.4.2 Batch Scheduling

Batch production is the manufacture of a finite number of identical items, to meet a specific order to satisfy continuous demand. The batch scheduling is generally done by using runout time method.

The runout time is that period of time for which previously scheduled production, plus inventory on hand, satisfies demands for an item. The basic objective of this method is to balance the utilization of production capacity, for example, machine hours, so that runout time for all items is the same. Production efforts are thus balanced across a group of items rather than concentrated on a few items. The runout method is explained with the help of following example:

Example 8.9: *The available machine hours for a facility in a productive system are 96.5 which are to be scheduled during a week. The productive facility is being used for 6 items A, B, C, D, E and F. The production time required for these items in machine hours, the inventory on hand in units, forecasted usage in units and forecasted weekly usage in machine hours is shown in Table 8.4*

Table 8.4

Col. No. *Item*	*1* *Inventory on Hand (units)*	*2* *Production time (machine hours per unit)*	*3* *Forecasted usage (units)*	*4* *Inventory on hand (in machine hours) (1 × 2)*	*5* *Forecasted weekly usage (machine hours) (2 × 3)*
A	125	0.2	60	25.0	12.00
B	250	0.08	85	20.0	6.80
C	75	0.5	30	37.50	15.00
D	300	0.09	96	27.00	8.64
E	239	0.15	78	35.85	11.70
F	98	0.7	42	68.60	29.40
Aggregate Total				213.95	83.54

Solution:

$$\text{Aggregate Runout time} = \frac{[\text{Aggregate inventory on hand (machine hrs)} + \text{Available machine hrs} - \text{Aggregate forecasted weekly usage}]}{\text{Aggregate forecasted weekly usage}}$$

$$= \frac{213.95 + 96.5 - 83.54}{83.54} = 2.72 \text{ weeks}$$

Based on a runout time of 2.72 weeks, the schedule required to achieve this runout time is shown in Table 8.5.

Table 8.5

Col. No. Item	*1 Desired Ending Inventory (Col. 3 of Table 8.4 × 2.72)*	*2 Total Items Required (Col. 3 of Table 8.4 + Col. 1 of Table 8.5)*	*3 Scheduled Units (Col. 2 of Table 8.5 – Col. 1 of Table 8.4)*	*4 Production machine h (Col. 3 of Table 8.5 × col. 2 of Table 8.4)*
A	163	223	98	19.6
B	231	316	66	5.3
C	82	112	37	18.5
D	261	357	57	5.0
E	212	290	51	7.7
F	114	156	58	40.6

Column 3 of Table 8.5 shows the number of units to be scheduled for items A, B, C, D, E and F in order to meet the constraint of available production hours as well as the inventory requirements.

8.4.3 Mass Production Scheduling

The scheduling approach used for mass production operations depends on the technology used. If the work is largely manual and utilizes a production line, as in the assembly stages of telephone manufacturing, the scheduling problem relates to determining the operatar task time required to achieve the desired output rate, and then evenly distributing these tasks among the production workers. These problems are referred to as line balancing problems. Apart from line balancing, mass production scheduling requires the coordination of flow of materials both to and from the point of the transformation. The management usually wants to avoid excessive inventory built-up at the starting point, at the various stages of manufacturing and finished goods inventory at the end of the process. At the same time, management also wants to ensure that raw materials, sub-assemblies and finished goods are available to ensure a smooth flow of production and to meet customer demand. Achieving this balance requires the development of sub schedules relating to the ordering of new materials, the completion time of sub-assemblies and the removal of finished products to storage areas.

8.5 LINE OF BALANCE (LOB) (OR LINE BALANCING)

The term Assembly line balancing is associated with the schedule of production line jobs that balances the workload of each work station so that each of the worker on the production line has to carry out more or less equal amount of work. The output of assembly line is dictated by the largest *station time* which becomes invariably *cycle time* of the line. The first step is to divde the whole work into elements and list them sequentially along with the time required to complete the element and group the tasks/ elements, which have to be performed into balanced work assignments. The objective is to group the tasks so that the output rate of each work station meets, as closely as, possible without exceeding, the time available for working on each unit, this procedure minimises that workers and equipment needed to operate the line at the desired rate. This process of grouping the tasks/elements is known as *line balancing*.

All work must be identified and broken down into elemetal tasks. These later are the minimal units of work, which cannot be further divided without the resulting division requiring significantly longer time to perform than the whole. Drilling a hole, for example, would be considered an elemental task here. If the hole were started on one drill press and completed on another, additional time would be required to withdraw the first drill and remove.the part from the press, to position the part on the second press, and to advance this drill to where it makes contact with the part—all the unnecessary activities, could be avoided if the task were completed on one machine.

Once all the elemental tasks are identified, the precedence table and precedence diagram are to be drawn. When writing precedence table, the factor to be remember is, the restrictions involved in doing a task also to be presented against the task. Now by means of some illustrated examples, let us discuss the methods used for balancing the line.

8.5.1 Heuristic Method

Heuristic means *searching to find out.* That is to find out (discover) things for oneself. Heuristic describes a particular approach to problem solving, decision-making and control. Heuristic are often simple, *thumb rule* which are used to solve complex problems. They help to provide ways to solve problems while at the same time beginning of an investigation, avoid rigorous logical analysis. Heuristic models utilise common sense, logic and more than all experience and common sense to tackle new problems. Heuristic methods, though, have little, if any, theoretical, yet they provide most likely (though not optimal) solutions, which are good enough from a practical point of view. Though we have many methods, we use Heuristic method for intermittent flow production system.

Steps to be followed in Heuristic Method:

(*a*) Identify the work (job/task).

(*b*) Breakdown the work into element tasks or steps.

(*c*) List the various elements along with their precedence relationship or logical relationships and the time required.

(*d*) Sketch the precedence diagram.

(*e*) Consider the highest time element in the table. This time will become *cycle time.*

(*f*) Add up all the elemental times and find the *total time.*

(*g*) Divide the total time by cycle time to get the number of work stations required.

(*h*) Assign task to stations or group the elements, so that each group is considered as a station. Here, we must care to see that the precedence relationship is not violated. Also total time of all the elements in a group does not exceed cycle time.

Example 8.10: *There are eight elements in completing a job. The precedence relationship and the time required (in minutes) to complete each element is given in the table. Draw LOB.*

Steps or elemental tasks	*Immediate predecessor*	*Duration of the element in minutes*
1	–	*3*
2	–	*4*
3	*1*	*2*
4	*2*	*5*
5	*3*	*4*
6	*5*	*8*
7	*4*	*2*
8	*6*	*4*
9	*8*	*6*
–	***Total:***	***38 minutes***

Solution: Total time = 38 minutes.

Highest time in the table = 8 minutes. Hence cycle time = 8 min.

Number of workstations = $\dfrac{\text{Total time}}{\text{Cycle time}} = \dfrac{38}{8}$ = 4.75, say approximately 5 stations.

(*a*) First we shall consider what happens if we take cycle time = total time. The line diagram will be as shown in Fig. 8.8 for the given problem.

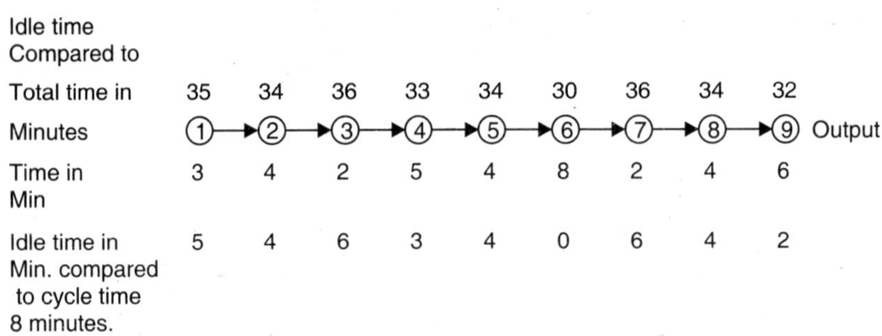

Fig. 8.8 Line diagram for the problem.

There are nine persons on the line, one for each station. Worker on station number 6 works for 8 min., compared to this the idle time for other workers are shown in the figure. Now for every 38 minutes one job is completed. Compared to the total time, idle time to each worker is also shown in the figure. If we consider that the span of working day as 8 hours,

$$\text{Productivity for the day} = \frac{(8 \times 60)}{38} = \frac{480}{38} = 12.6 \text{ jobs.}$$

This is to say that 12.6 jobs will be completed per day with nine workers on the line.

We see that most of the time the workers are idle and the organisation has to pay for the idle time. Hence the labour productivity is very low and the product cost is high.

Suppose, we group the elements in such a way that the total time of the group does not exceed cycle time (grouping should be done without violating the precedence relationship given), we will get a better-balanced line. Figure 8.9 shows the precedence diagram.

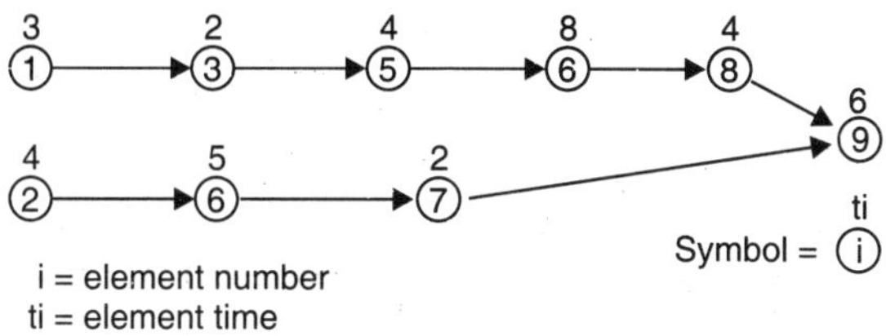

Fig. 8.9 Precedence diagram

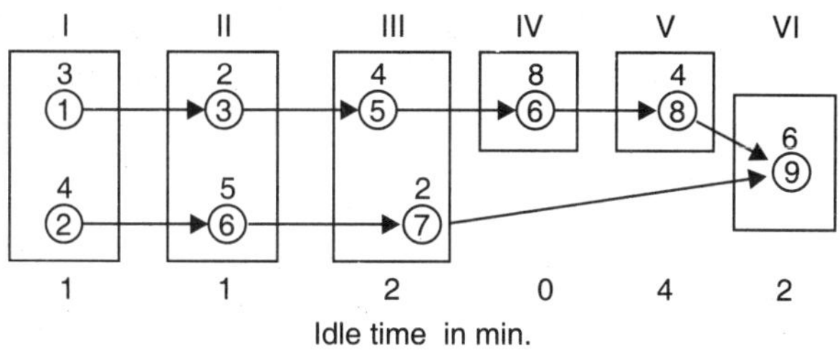

Fig. 8.10 Grouping of elements so that the total time of each group does not exceed 8 min.

Figure 8.11 shows the line of balance for cycle time = 8 minutes. Here we have 6 stations. Station V has idle time of 4 minutes and station VI has 2 minutes of idle time. Stations II and I have time of 1 minute each. Station III has idle time of 2 minutes. Idle time of station IV is zero minutes.

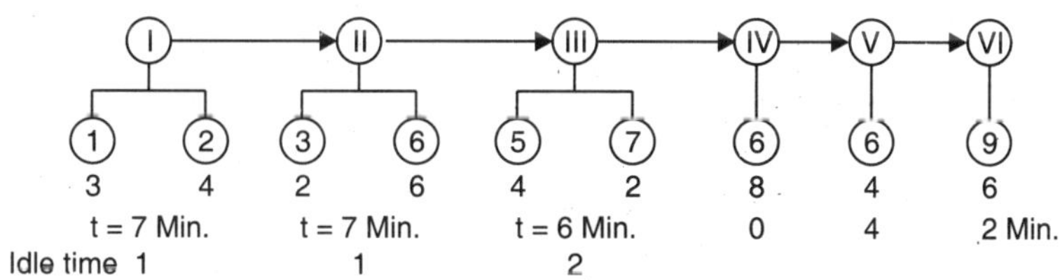

Fig. 8.11 Line of balance for cycle time = 8 minutes.

Daily productivity = (8 × 60)/8 = 60 units per day (with 6 workers only). As we see uneven idle time, the line is not perfectly balanced. The uneven idle time may lead to number of labour problems.

Solution (*ii*): We shall try to write a LOB with a cycle time = 10 minutes. Figures 8.12 and 8.13 shows the grouping of elements and the line of balance respectively.

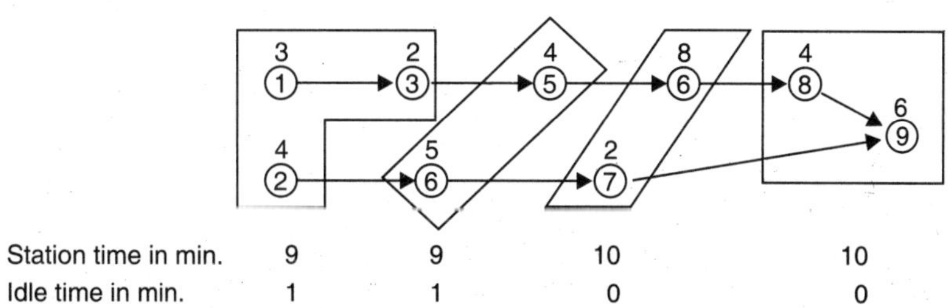

Fig. 8.12 Grouping of elements for cycle time = 10 minutes.

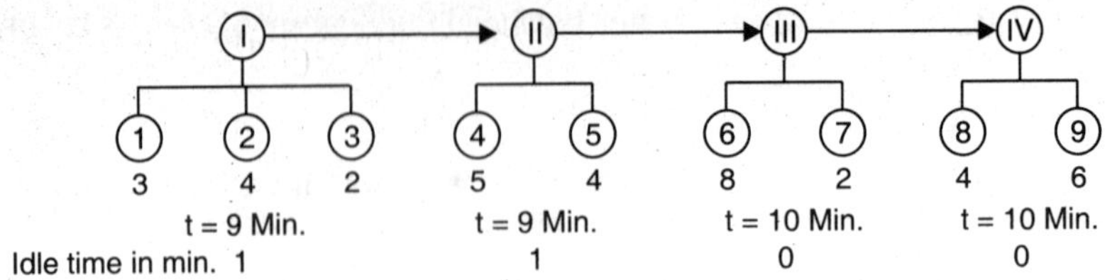

Fig. 8.13 Line of balance to cycled time = 10 minutes

Total time = 38 min. Cycle time = 10 min.

Number of stations = $\frac{38}{10}$ = 3.8, say approximately 4 stations.

Taking 8 hour working day, daily productivity = $\frac{(60 \times 8)}{10}$ = 48 jobs per day.

There are four stations and idle times at station II and I are 1 minute each. Hence, the line is better balanced.

In case we try the same problem with cycle time as 12 minutes, then also we get an unbalanced line, with too much of variation in station idle times. While selecting cycle time and grouping of items, the following points are to be remembered.

Depending upon the desired production rate of the line, the cycle time (CT) or the time btween the completions of two successive assemblies can be determined. This dictates the conveyor speed in the assembly line or the time allocated to each operator to complete his share of work on a manual line. The individual work elements or tasks are then grouped into work stations such that:

(*i*) The station time (ST), which is the sum of the times of work elements performed at the station and should not exceed the cycle time (CT).

(*ii*) The procedure restrictions implied by the precedence diagram are not violated.

There are many possible ways to group these tasks keeping the above restrictions in mind, we can use line efficiency (LE), balance delay (BD), and smoothness index (SI) to measure how good or bad a particular grouping is. This is explained below:

(*a*) ***Line Efficiency (LE):*** This is the ratio of total station time to the product of cycle time and the number of workstations.

L.E = $\frac{\Sigma ST_i}{K \times CT}$ × 100, where ST = station time of station '*i*', K = total number of stations and CT = cycle time.

In the example solved above cycle time = 10 mins. Hence

$$L.E. = \frac{9+9+10+10}{4 \times 10} = \frac{38}{40} = 0.95, \text{ hence line efficiency} = 95\%.$$

(*b*) ***Balance Delay (BD):*** This is the measure of the time efficiency and is the total idle time of all stations as a percentage of total available working time of all stations.

$$B.D = \frac{\{K(CT) - \Sigma ST_i\}}{\{K \times (CT)\}} \times 100, \text{ For the solved example,}$$

$$B.D = \frac{\{(4 \times 10) - (9+9+10+10)\}}{4 \times 10} = \frac{(40-38)}{40} = \frac{2}{40} = 0.05 = 5\% \text{ OR}$$

Balance delay = 100 – LE as a percentage. Here it is 100 – 95 = 5%.

(*c*) ***Smoothness Index:*** This is an index to indicate the relative smoothness of a given assembly line balance. A smoothness index of zero (= 0) indicates perfect balance. This is given by:

$SI = \sqrt{\Sigma(ST_{max} - ST_i)^2}$ Where ST_{max} = Maximum station time, ST_i = Station time of station '*i*', K = Total number of stations.

For the solved example,

$$SI = \sqrt{\{(10-9)^2 + (10-9)^2 + (10-10)^2 + (10-10)^2\}}$$

$$= \sqrt{1+1+0+0} = \sqrt{2} = 1.4.$$

It may be noted that in designing an assembly line the number of work stations, K cannot exceed the total number of elements, N. (In fact, K is an integer such that $1 \leq K \leq N$). Also the cycle time is greater than or equal to the maximum time of any work element and less than the total of work element times at that station. That is $T_{max} \leq CT \leq \Sigma\, T_i$,
where, T_i = Time of work element '1', N = Total number of work elements, T_{max} = Maximum work element time and CT = Cycle time.

There is yet no satisfactory methodology, which guarantees optimal solution for all assembly line balancing problem. The emphasis has been on the use of heuristic method that can obtain a fairly good line balance for the given problem.

8.5.2 Kilbridge and Wester Method of Balancing Assembly Line

Kilbridge and Wester propose this method. Here numbers are assigned to each operation describing how many predecessor it has. Operations with the lowest predecessor number are assigned first to the work stations. The steps to be followed in this method are explained by means of an example.

Example 8.11: *A task having 12 work elements and their logical relationship and time (in seconds) taken is given in the table below. Draw the line of balance.*

Element No.	*Predecessor element*	*Time in seconds*
1	–	5
2	1	3
3	2	4
4	1	3
5	4	6
6	3, 5	5
7	6	2
8	7	6
9	9	1
10	6	4
11	10	4
12	8, 9, 11	7
–	Total time	50 seconds

Solution: Step 1. Construct the precedence diagram (Fig. 8.14) where, i = element number, j = time in seconds = i

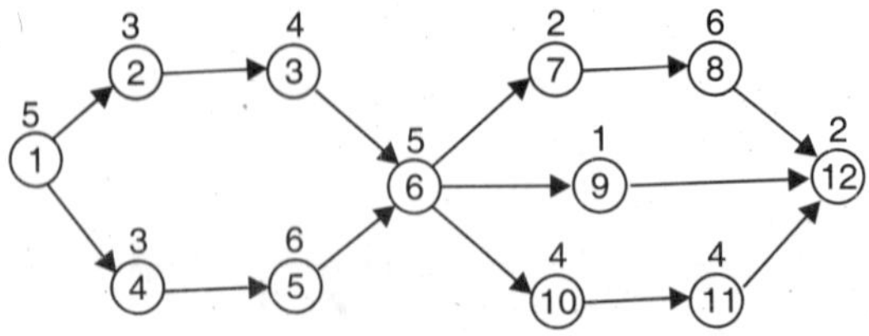

Fig. 8.14 Precedence diagram.

Step 2. In precedence diagram identify the element which do not have predecessors (does not follow other element) and enclose in a column and number that column as I. Next identify the element those immediately follow the element in column number I and enclose them in column number II. Similarly in column number III enclose all elements that follow the elements in column number II. Continue this procedure until all elements are enclosed in columns. The element in the last column will not have more elements to follow. By so constructing the columns, the elements within a column can be assigned to works stations in any order, provided all the elements of the previous column have been assigned. At the same time we should not violate precedence order.

Step 3. Select a feasible cycle time (CT). By a feasible time we mean one for which $T_{max} \leq CT \leq ST_i$

Step 4. Assign work elements to the station, such that the sum of elemental times at a station does not exceed the cycle time, CT. This assignment precedes from column I to II and so on, breaking intra column ties using the criterion of minimum number of predecessors.

Step 5. Delete the assigned elements from the total number of work elements and repeat the above step until all the elements are assigned.

Step 6. If the station time exceeds the cycle time, due to the inclusion of a certain work element, this work element should be assigned to the next station.

Repeat steps 4 to 6 until all elements are assigned to work stations.

In the given problem there are 12 elements. They are arranged in columns as shown in Fig. 8.15. The same is given in the Table 8.8.

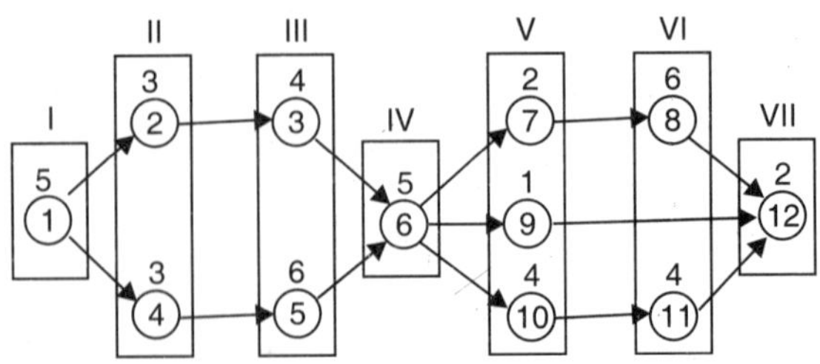

Fig. 8.15 Precedence diagram with columns.

Table 8.8: Showing arrangement of elements in the columns

Elements	1	2, 4	3, 5	6	7, 9, 10	8, 11	12
Columns	I	II	III	IV	V	VI	VII

Let us take cycle time (CT) = 10 seconds. Let us count the number of predecessor for each work element and record in Table 8.7. Work element number 1 is selected first becasue it has no predecessors (or least number of predecessors). So we assign element to station number I. Either element 2 or 4 each of which has an operation time of 3 seconds, can be assigned to station I, which results in a station time of 8 seconds, which is less than cycle time (=10 seconds). If we allocate element 2 to station I, element 4 cannot be allocated to station I, as station time will be greater than cycle time. Allocating the element 4 to station II and we continue on the same logic, until all elements are allocated, the allocation will be as shown in Table 8.9 and Figure 8.16.

Table 8.9: Number of predecessors for each element

Work element (i)	*Number of Predecessors*	*Time (T_i) in seconds*
1	0	5
2	1	3
3	2	4
4	1	3
5	2	5
6	5	5
7	7	1
8	7	6
9	6	1
10	6	4
11	7	4
12	11	7

Table 8.10: Assignment of work element to stations

Station number	*Element (i)*	*Time T_i in seconds*	*Station time in seconds*	*idle time in seconds (Cycle Time = 10 sec.)*
I	1, 2	5 + 3	8	2
II	4, 5	3 + 6	9	1
III	3, 6	4 + 5	9	1
IV	7, 9, 10	2 + 1 + 4	7	3
V	8, 11	6 + 4	10	0
VI	12	7	7	3

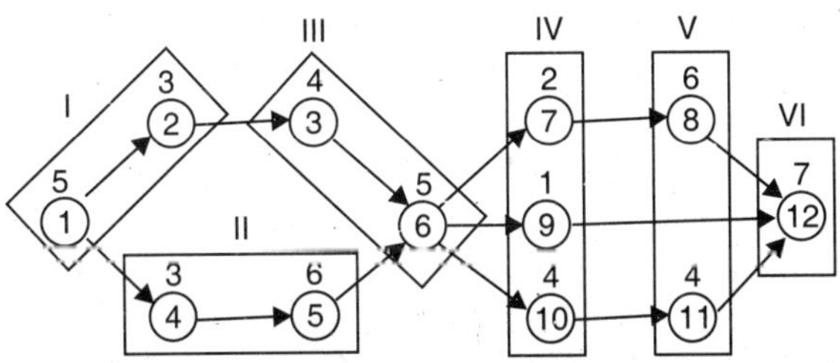

Fig. 8.16 Allocation of elements to stations.

$$\text{Line efficiency} = \text{LE} = \left\{\frac{50}{(6\times10)}\right\} \times 100 = 83.3\%$$

$$\text{Balance delay} = \text{BD} = 100 - 83.3 = 16.7\%$$

$$\text{Smoothening index} = \text{SI} = \sqrt{(4+1+1+9+9)} = \sqrt{24} = 4.89.$$

As the solution obtained above is not an optimal because smoothening index is 4.89. Now let us take cycle time = 9 seconds and balance the line. The allocation will be shown in Table 8.11.

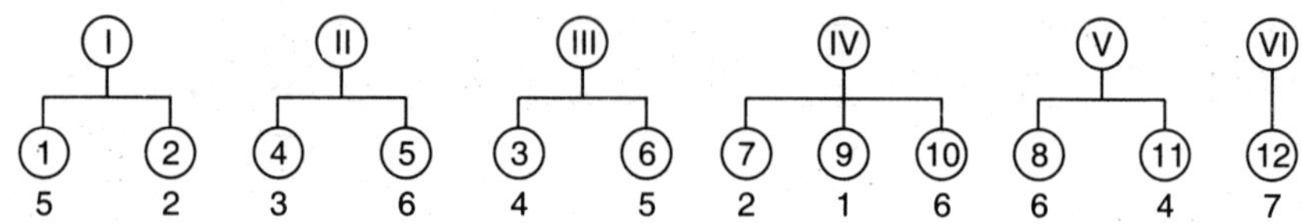

Fig. 8.17 Line of Balance

Table 8.11: Allocations of elements to stations with cycle time = 9 seconds

Station	*Element*	*T_i in in seconds*	*Station time in seconds*	*idle time when CT = 9 sec.*
I	1, 2	5 + 3	8	1
II	4, 5	3 + 6	9	0
III	3, 6	4 + 5	9	0
IV	7, 8	2 + 6	8	1
V	10, 11	4 + 4	8	1
VI	9, 12	1 + 7	8	1

$$\text{Line efficiency} = \text{LE} = \frac{50}{(6\times9)} \times 100 = 92.6\%$$

$$\text{Smoothening Index} = \text{SI} = \sqrt{(1+1+1+1)} = \sqrt{2} = 2$$

This should cause no concern, for the Kilbridge and Wester method. As the method is heuristic method, which gives us a good working solution, with little computational effort.

8.6 AGGREGATE PLANNING

Production planning in the intermediate range of time is termed as '*Aggregate Planning*'. It is thus called because the demand on facilities and available capacities is specified in aggregate quantities. For example, aggregate quantities of number of automobile vehicles, aggregate number of soaps, etc. Here the total expected demand is specified without regard to the product mix that makes up the specified figure.

While dealing with production problems, the *planning process* is normally divided in three categories. (*i*) *Long range planning* which deals with strategic decisions such as purchase of facilities, introduction of new products, processes, etc., (*ii*) *Short-term planning,* which deals with day-to-day work, scheduling and sometimes inventory problems (*iii*) *Intermediate planning or aggregate planning,* which is in between long range and short-term planning, which is concerned in generally acceptable planning taking the load on hand and the facilities available into considerations. In aggregate planning the management formulates a general strategy by which capacity can be made to satisfy demand in a most economical way during a specific moderate time period, say for one year. The aggregate planning is made operational through a *master schedule* that gives the manufacturing schedule (products and dates of manufacture), generally, day-to-day schedules are prepared from master shedule. Facility planning and scheduling has got very close relationship with aggregate planning. Figures 8.18 and 8.19 shows aggregate planning and its relationship with other planning.

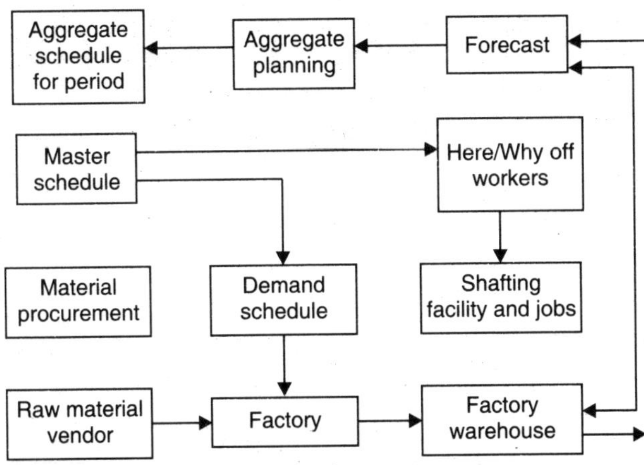

Fig. 8.18 Aggregate planning and production system.

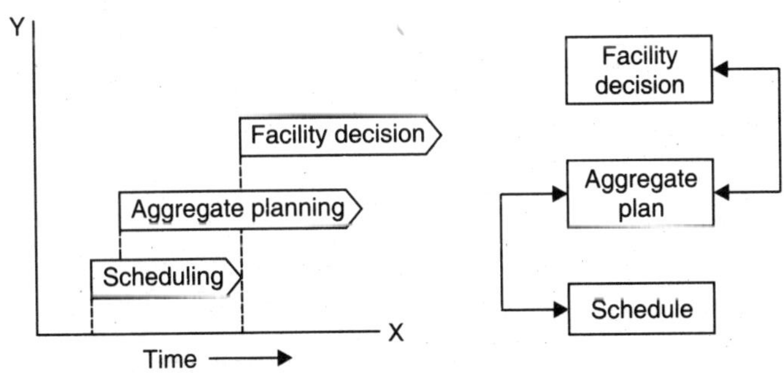

Fig. 8.19 Relationship between facility planning, aggregate planning and scheduling.

Properties of aggregate planning: To facilitate the production manager the aggregate planning must have the following characteristics:

(*i*) Both output and sales should be expressed in a logical overall unit of measuring. For example, an automobile manufacturing can say 1000 vehicles per year, without giving the number of each varity of vehicle. Similarly a paint industry can say 10,000 liters of paint and does not mention the quantities of each colour.

(*ii*) Acceptable forecast for some reasonable planning period, say one year.

(*iii*) A method of identification and fixing the relevant costs associated with the plant.

(*iv*) Availability of alternatives for meeting the objective of the organisation.

(*v*) Ability to construct a model that will permit to take optimal or near optimal decisions for the sequence of planning periods in the planning horizon.

(*vi*) Facilities that are considered fixed to carry out the objective.

8.6.1 Decision Options in Aggregate Planning

Many a time the production manager may face a situation, where sometimes he finds that the demand for the product or forecasts made is very much less than the normal available capacity or the forecast may be very higher than the available capacity. In such cases he has to find a way out to match these two extremes. For him *decision options* are made available to modify demand or supply. The modifications are normally followed by the development of a few specific decision rules to plan aggregate output for both manufacturing and service industries. The decision options are:

(*i*) The decisions that modify the demand for products and services.

(*ii*) The decisions that modify the supply of product and services to match the demand.

(*i*) ***Modification of demand:*** The demand for a product or service may be increased or decreased by adopting various strategies, discussed below:

(*a*) *Differential pricing:* If the available capacity is higher than the demand for the product, then to increase the demand the company may announce the discounts for the purchases made during certain period. You might have heard about off-season rates, which are much lesser than that of seasonal rates, like this we can get good number of examples, to increase the demand. Similarly, if the demand is higher than the available capacity during a period, to reduce the demand for the product/service differential prices are fixed for the product so that the consumers may postpone their desire to purchase the product, when the price is reduced, *i.e.*, during off-season.

(*b*) *Advertising and sales promotion:* To smoothen the demand to match the supply the advertisement is so timed to see that demand is increased in slack period or to shift the demand from peak period to slack period. By this the production manager is able to match the demand with the production capacity.

(*c*) *Backlogs:* To shift the demand from peak period to slack period, manufacturer can create backlogs and request the customer to wait for delivery.

(*d*) *Development of complimentary products:* Manufacturers of products, which are highly seasonal in nature, can adopt this strategy. Here a complimentary product is manufactured by using the available facility and skill so that the plant is kept engaged in manufacturing activity. By doing so they can discontinue the manufacture of the regular product to avoid building up of inventory.

(*ii*) ***Modification of supply:*** Some of the methods by which we can modify the supply to match the demand are:

(*a*) *Hiring and lay-off of employees:* As this policy deals with the employees, one must be careful enough while selecting this strategy. It is because by hiring labours on temporary basis, the hiring charges may go high or sometimes the workers may not available in time when needed. Or, if the manager goes for lay-off strategy, he is going to invite the trouble from the labour unions, at the same time the employee morale may go down. Sometimes the company may have to face legal problems due to lay-off. Before taking decision one has to weigh both positive and negative aspects so that the decision taken will not hurt anybody.

(*b*) *Overtime and under time:* When the changes in supply are meager and temporary, overtime and under time are most commonly used. By adopting overtime, we increase the production but overtime costs are to be added to the normal costs. By under time the production may be reduced and under time costs are to be considered here.

(*c*) *Use of part-time or temporary labour:* When the manager has to increase the supply, he has to increase the capacity. By employing the temporary or part-time labour the manager can do this. Depending on the type of labour the cost may be slightly higher or lower than the normal labour cost.

(*d*) *Subcontracting:* In case the other company has some spare capacity and is ready to hire the capacity, the manager can give a part of production as subcontract to that firm.

The parent company always can send its workers to have a close control over the quality of the product produced. Depending on the hire charges, the costs of products produced under subcontracting may slightly higher than the normal cost.

When preparing aggregate planning, the production manager can select any of the above-discussed strategies, but a combination, rather than a single strategy, will result in the most economical plan. A few mathematical methods are available to select the optimal combination of the strategies, such as linear programming model, Transportation model; non-linear programming model may be used.

8.6.2 Aggregate Planning Methods

No aggregate planning methods yield truly *optimal* rates of production. For example, graphic and charting technique basically works with few variables at a time on a trial and error basis. Some mathematical methods also follow this pattern, whereas others begin with limiting or unrealistic assumptions and achieve a theoretical optimality. The problem then remains to reconcile this with real-world problem. In fact, charting technique is simplest one and preferred by many managers. In the following paragraph let us discuss some of the methods.

Graphical and charting technique

Here we construct simple tables and graphs that enable the management to compare and contrast forecasted demand with existing capacity. First thing one has to do is to construct a cumulative production requirement chart for the planning period. Then the production manager has available capacity on one hand and the fluctuating demand on the other. He tries to fulfil the demand by various combinations of workforce, overtime/ under time, subcontracting and inventory maintenance and so on.

Example 8.12: *M/s XYZ Company has made the forecast given below and now it has to plan its production so as to satisfy the demand. Required safety stock and the number of production days in a month is given below. In addition the following assumptions are made as far as the resources available with the company and other production alternatives to go for aggregate planning.*

(*i*) *Normal production capacity is 700 units per day.*

(*ii*) *At most 900 units may be produced per day by allowing overtime at an extra cost of ₹15 per unit.*

(*iii*) *In case the production requirement is greater than 900 then subcontracting can be used at an extra cost of ₹20 per unit.*

(*iv*) *Cost of hiring a person is ₹1000 and cost of lay-off is ₹2000.*

(*v*) *Production rate per person is 100 units per day.*

(*vi*) *Number of persons working is in production shop is 5 and average salary per month is ₹800 per month.*

(*vii*) *Cost of carrying inventory is ₹20 per unit per year.*

(*viii*) *Opening stock is 5000 units.*

Solution: Using graphical method, first we have to. construct cumulative production requirement. We can get Cumulative production requirement = (Cumulative demand + Safety stock requirement for the month – Opening stock at the beginning of the planning period).

In the graph (Fig. 8.20), the line A represents the cumulative production requirement line. Line B and C represents two alternative production plan to meet the production requirement. The third alternative is same with the line A that is the production plan is same as actual production requirement. These three alternatives are as under.

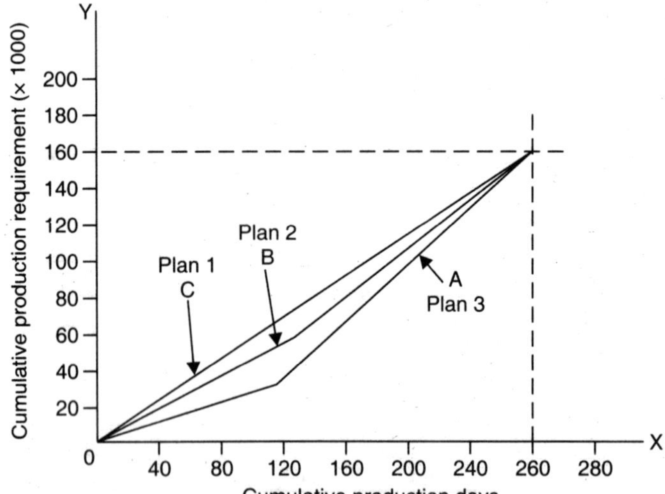

Fig. 8.20 Cumulative demand and production.

Table 8.12: Forecast of sales and safety stock of M/S XYZ Company

Month	*Expected demand*	*Required safety stock*	*Productive days*
Jan.	9000	4000	21
Feb.	6000	2600	19
March	5000	2500	25
April	6000	2800	22
May	12000	5000	22
June	16000	7000	23
July	18000	8000	20
August	20000	9000	20
Sept.	16000	8000	22
Oct.	18000	8000	22
Nov.	14000	6000	16
Dec.	16000	7000	20
Total	**156000**	**69900**	**252**

Table 8.13: Cumulative production requirement

Months	*Cumulative demand*	*Cumulative production requirement*	*Cumulative production days*
January	9000	8000	21
February	15000	12600	40
March	20000	17500	65
April	26000	23800	87
May	38000	38000	109
June	54000	56000	132
July	72000	75000	152
August	92000	96000	172
September	108000	111000	194
October	126000	129000	216
November	140000	141000	232
December	156000	158000	252

(Closing stock is 5000 units)

1. The production rate is 627 unit average per day with six people working throughout the year. The workers with an extra cost of ₹ 400 per day may complete extra 27 per day. (158000/252 App. = 627) and (27 × 15 = ₹ 405).
2. Production for first five months is at the rate of 459 units per day with five persons and the production rate for next seven months is at a rate of 755 units

per day with 7 people working and earning of overtime is considered. Hiring of two persons at the beginning of six months is considered.

3. Production rate is taken on actual requirement per month.

We can consider many more alternatives to reach the desired goal. But to explain the procedure, the above three plans are given.

Table 8.14: Cost calculations for three possible plans

Plan 1:

Month	*Production plan*		*Cumulative production plan*	*Cumulative production requirement*	*End of month seasonal inventory*
	Per day	*Per month*			
January	627	13167	13167	8000	5167
February	627	11913	25080	12800	12280
March	627	15675	40755	17500	23255
April	627	13794	54549	23800	30749
May	627	13794	68343	38000	30343
June	627	14421	82761	56000	26761
July	627	12540	95304	75000	20304
August	627	12540	107844	96000	11844
September	627	13794	121638	108000	13638
October	627	13794	135432	129000	6432
November	627	10032	145464	141000	4464
December	627	12540	158004	158000	(4)
Total					**185237**

$$\text{Average seasonal anticipation stock} = \frac{185237}{12} = 15436$$

$$\text{Average safety stock} = \frac{69900}{12} = 5825$$

Cost of carrying seasonal inventory = 15436 × ₹ 20 = ₹ 308720

Wages = 6 × 800 × 12 = ₹ 57600

Incentive for producing extra 27 units = 400 × 252 = ₹ 100800

Hiring charges for one person = 1 × 1000 = ₹ 1000 (six persons employed in planning period)

Total cost for the plan number one

= ₹ 308720 + ₹ 57600 + ₹ 100800 + ₹ 1000

= ₹ 468120

Table 8.15: Cost calculations for three possible plans

Plan 2:

Month	*Production plan*		*Cumulative production plan*	*Cumulative production requirement*	*End of month inventory*
	Per day	*Per month*			
January	459	9639	9639	8000	1639
February	459	8721	18360	12800	5860
March	459	11475	29835	17500	12335
April	459	10098	39933	23800	161 33
May	459	10098	40031	38000	12031
June	755	17365	67396	56000	11396
July	755	15100	82496	75000	7496
August	755	15100	97596	96000	1596
September	755	16610	114206	108000	6206
October	755	16610	130816	129000	1816
November	755	12080	142896	141000	1896
December	755	15100	157996	158000	(–4)
Total					**78104**

$$\text{Average seasonal anticipation stock} = \frac{78104}{12} = 6509 \text{ units.}$$

$$\text{Average safety stock} = \frac{69900}{12} = 5825$$

Cost of carrying seasonal inventory = 6509 × ₹ 20 = ₹ 130180

Wages = (5 × 5 + 7 × 7) × 800 = ₹ 59200

Overtime = (55 Nos. × 143 days × ₹ 15 per unit) = ₹ 117975

Hiring charges = ₹ 2000 (Hiring two persons more at the beginning of June)

Total = ₹ 130180 + 59200 + 117975 + 2000 = ₹ 309355

Table 8.16: Cost calculations for three possible plans

Plan 3:

Month	*Average production per/day required*	*Number of workers*	*Wage*	*Number of persons hired*	*Number of persons layed off*	*Cost of hiring and laying off*	*Cost of over time*	*Cost of subcont-racting*
January	381	4	3200	–	1	2000	–	–
February	253	3	2400	–	1	2000	–	–
March	188	2	1600	–	1	2000	–	–
April	286	3	2400	1	–	1000	–	–
May	675	7	5600	4	–	4000	–	–
June	783	7	5600	–	–	–	28635	
July	950	7	5600	–	–	–	60000	20000
August	1050	7	5600	–	–	–	60000	60000
September	636	7	5600	–	–	–	–	–
October	864	7	5600	–	–	–	54120	–
November	750	7	5600	–	–	–	12000	–
December	850	7	5600	–	–	–	45000	–
Total			**54400**			**11000**	**259755**	**80000**

Average seasonal anticipation stock = 0

Average safety stock = 5758

Wages = ₹ 54400

Cost of hiring and laying off = ₹ 11000

Cost of overtime = ₹ 259755

Cost of subcontracting = ₹ 80000

Total cost = ₹ 54400 + 11000 + 259755 + 80000 = ₹ 405155

Before drawing a graph, let us compare the costs of all the three plans. Table 8.17 shows the cost comparison of the entire three plans.

Tables 8.17: Cost comparison of three plans

Item	*Plan No. 1*	*Plan No. 2*	*Plan No. 3*
Seasonal inventory	308720	130180	0
Wages	158400	59200	54400
Hiring and lay-off charges	1000	2000	11000
Overtime	–	117975	259755
Sub contracting	–	–	80000
Total ₹	**468120**	**309355**	**405155**

***Note*:** The cost of safety stock (₹ 20 × 5758 = ₹ 115160) is same for all the plans and hence not shown in the table. Depending on the need and available alternatives different combination plans can be prepared and the best one is selected. In the above table, we can see that plan 2 is best one.

Aggregate Planning by Mathematical Techniques

Operations research models may be used for solving aggregate planning problems. In addition some scholars have developed their own methods for solving aggregate planning problems.

(*a*) **Aggregate planning by linear programming:** Vajda (Planning by Mathematics, 1973) has dealt with simple linear programming model based on transportation problem to solve aggregate planning problem. He has taken a problem of a producer who is required to supply a product of specified number of customers for future period. Cost of overtime and regular production are given and capacities of production by regular and overtime are specified. The problem is to make optimal allocation.

Example 8.13: *A manufacturer can produce 9 units in regular hours and 3 units by overtime, production requirement for four months are:*

Month 1	*Month 2*	*Month 3*	*Month 4*
5	6	8	6

Cost of production during regular and overtime production is:

Regular production:	₹	*1*	*4*	*2*	*4*
Overtime production:	₹	*2*	*6*	*4*	*6*

Shortage cost is assumed to be 1 unit cost per unit per month. The cost matrix is given below:

Symbols used in the matrix: R = Regular production, O = Overtime production, M = High cost factor, so that the cell having M as the cost should not be allocated.

Dummy column is to balance the rim requirements.

Month		1	2	3	4	Dummy	Capacity
1	R	1	2	3	4	0	9
	O	2	3	4	5	0	3
2	R	M	4	5	6	0	9
	O	M	6	7	8	0	3
3	R	M	M	2	3	0	9
	O	M	M	4	5	0	3
4	R	M	M	M	4	0	9
	O	M	M	M	6	0	3
Requirement:		5	6	8	6	23	

Cost Matrix of prodeuction planning system.

Solution:

Month	Availability	1	2	3	4	Dummy	
1	R	5	4	–	–	0	9
	O	–	2	–	–	1	3
2	R	–	–	–	–	9	9
	O	–	–	–	–	3	3
3	R	–	–	8	1	0	9
	O	–	–	–	–	3	3
4	R	–	–	–	5	4	9
	O	–	–	–		3	3
Requirement:		5	6	8	6	23	

The production plan is as follows:

(*i*) In the first month, the production quantities in regular time and in overtime are 9 and 2 respectively. The production requirement of first month is completely catered and five units of inventory is carried over to the second month.

(*ii*) In the second month, no production is made and the demand of the second month is met with the inventory carried over for the first month.

(*iii*) In the third month, the production of 9 units is made in regular time only. With this, production requirement of third month is fully met and one unit of production is carried for the fourth month.

(*iv*) In the fourth month, 5 units are produced in regular time to satisfy the demand of fourth month.

(*b*) **Linear decision rule (LDR):** Mr. Holt, Mr. Modigliani, and Mr. Muth of Carnegie Institute of Technology have developed a special type of mathematical technique known as linear decisions rule to solve production planning problem. This technique deals with quadratic cost function for the organisation with cost components such as *regular payroll, hiring and layoff costs, overtime cost and inventory carrying cost.*

Each of the actual cost curves is approximated to some theoretical cost curves (thin lines) as shown in Fig. 8.21. The cost curves in Fig. 8.21 are assumed to be quadratic and as such the total cost curve contains three squared items. Figure 8.21(A) shows the cost function of regular payroll. The work size in the model is adjusted once per period and it has been assumed that cost due to the workforce varies proportional to the workforce size. Figure 8.21(B) shows the hiring and lay off cost. Figure 8.21(C) shows the cost of overtime and under time for various production rates, keeping workforce size as constant. If the production rate varies, the management may opt for overtime or under time or else they may hire or lay off workers. As the management will be interested in minimising the cost of production, its decision depends on the combination of options, which minimises the total production cost. Cost of holding inventory is shown in Fig. 8.21(D). The model is:

Minimise the total cost, C_t = Regular payroll cost + Hiring and Lay off costs + Overtime Cost + Inventory Cost

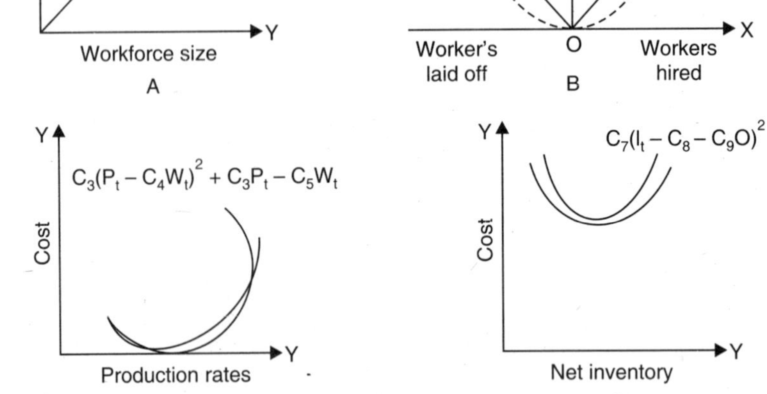

Fig. 8.21 (A, B, C and D): Quadratic cost Assumptions (thin lines) in linear decision rule.

This expression is quadratic and if it is differentiated with respect to workforce size and also with repsect to production rate, two linear equations are generated. These equations are to be solved to get the minimum point on the total cost curve.

Example 8.14: *The table given below gives the data for LDR problem. Assume that the beginning inventory is f_{t-1} is 90 units and the present workforce W_{t-1} is 28 workers. Calculate workforce size and production rate.*

Table 8.17: Data for LDR calculations

Month	*Forecast* 1	*Production Coefficient* 2	*Equation Forecast effect* 3 = 1 × 2	*Workforce coefficient* 4	*Equation forecast effect* 5 = 1 × 4
January	120	0.463	55.56	0.0101	1.212
February	125	0.234	29.25	0.0088	1.1
March	135	0.111	24.99	0.0071	0.958
April	145	0.046	6.67	0.0054	0.783
May	155	0.013	2.15	0.0042	0.651
June	160	–0.002	–0.32	0.0031	0.496
July	165	–0.008	–1.32	0.0023	0.38
August	150	–0.01	–1.5	0.0016	0.24
September	140	–0.809	–1.26	0.0012	0.168
October	135	–0.008	–1.08	0.0009	0.122
November	125	–0.007	–0.87	0.0006	0.075
December	120	–0.005	–0.6	0.0005	0.06
Total			**101.57**		**6.245**

Solution: If P_t = production rate for the period 't'

$W_{t\ 1}$ = Size of the workforce for the period $(t-1)$,

I_{t-1} = Inventory holding for the period $(t-1)$,

O_t = Forecasted production rate for the period 't'

$$P_t = \begin{bmatrix} +0.463\ O_t \\ +0.234\ O_{t+1} \\ +0.111\ O_{t+2} \\ +0.046\ O_{t+3} \\ +0.013\ O_{t+4} \\ -0.002\ O_{t+5} \\ -0.008\ O_{t+6} \\ -0.010\ O_{t+7} \\ -0.009\ O_{t+8} \\ 0.008\ O_{t+9} \\ -0.007\ O_{t+10} \\ -0.005\ O_{t+11} \end{bmatrix} + 0.993\ W_{t-1} + 153 - 0.464\ I_{t-1} \qquad \ldots(8.1)$$

$$W_t = 0.743\, W_{t-1} + 2.09 - 0.010\, I_{t-1} + \begin{bmatrix} +0.0101\, O_t \\ +0.0088\, O_{t+1} \\ +0.0071\, O_{t+2} \\ +0.0054\, O_{t+3} \\ +0.0042\, O_{t+4} \\ +0.0031\, O_{t+5} \\ +0.0023\, O_{t+6} \\ +0.0016\, O_{t+7} \\ +0.0012\, O_{t+8} \\ +0.0009\, O_{t+9} \\ +0.0006\, O_{t+10} \\ +0.0005\, O_{t+11} \end{bmatrix} \quad \ldots(8.2)$$

In the above equation first part of the right hand side of first equation and last part of the right hand side of second equation is the forecasted effect of the planning period where planning period is taken as twelve months.

Substituting the values in the first equation and simplifying, we get

$P_t = 101.57 + 0.993\ (28) + 153 - 0.464\ (90) = 240.61.$

Substituting the values in equation 2, we have

$W_t = 0.743\ (28) + 2.09 - 0.01\ (90) + 6.245 = 20.139.$

(*c*) **Heuristic method:** These methods are thumb rule methods. They do not give optimal solution but generally provide a workable solution for planning the production. We have 1. management coefficient method and 2. Search decision rule.

1. ***Management coefficient method:*** Models discussed above attempts to capture the important trade-offs required by complex analytical structures. Difficulties arise, however, when we estimate the costs required for complex mathematical models. As an alternative, *Bowman* developed the management coefficient model on the premise that management's past decisions can be incorporated into a system for improving present decision. This involves rigorous statistical analysis of management's past decisions. This method assumes that management is quite aware of and sensitive to the criteria of the system, managers then make a decision through a process of natural screening with a sense and intuition that may relate to the system variables even if the decision is more erratic than biased. It has been seen that many a times even the manager is not able to explain the reason for taking such a decision. Possibly the decision is taken by him subconsciously after seeing and experiencing a number of small occurrences.

Bowman's theory states that the management's performance can considerably improved upon by applying these rules consistently. One of the reasons that a set of decision rules will yield better results is due to the fact that most of the cost functions are shallow and dish shaped at the bottom and any decision that is slightly deviated from the optimum decision does not change the cost criteria drastically from the value that corresponds to the optimum solution.

Bowman in his model developed a procedure in which a multiple regression analysis of management's past behaviour on a particular decision set is performed. The coefficient so derived can be used to indicate any future decision of such type. The decision, it is expected, will still have some amount of business but the variance can be minimised.

2. Search decision rule (SDR): Search decision rule (SDR) is another type of the heuristic model that can be used for solving production-planning problems.

This model takes into account two linear feedback decision rules. The first one is workforce size, which includes two parameters: (*a*) work-force changes and (*b*) overtime. The second one is production rate, which includes two more parameters: (*a*) Back orders and (*b*) Inventory. These four parameters in various combinations generate numerous alternatives for planning purpose. A heuristic search procedure may then be applied to get the values of four parameters that can result the minimum cost of operation. There are other SDR models developed by *Jones and Tauberi.* Jones model is *parametric production planning model* in which the parameters for four cost functions are assumed to be quadratic in nature. Tauberi's SDR model assumes cost functions either quadratic or other mathematical expressions. In this model, cost functions are taken as best estimates of actuals. Another model developed by *virgin* uses simulation approach, where cost functions can be taken as linear, quadratic or step functions.

EXERCISES

1. Define loading, scheduling and sequencing of operations.
2. Five jobs A, B, C, D and E are to be assigned to 5 machines 1, 2, 3, 4 and 5. The costs associated with the manufacturing of each of these jobs on these five machines are shown in the following matrix. Determine the optimum assignment so as to minimise the total cost and calculate the total cost of this assignment.

Machines			*Jobs*		
	A	*B*	*C*	*D*	*E*
1	10	5	9	18	11
2	13	19	6	12	14
3	3	2	4	4	5
4	18	9	12	17	15
5	11	6	14	19	10

3. Describe briefly the various forms of scheduling? State the various measures to reduce the problems related to scheduling.
4. Explain job scheduling, mass and batch scheduling.
5. Explain line balancing.
6. Explain aggregate planning methods.

❑❑❑

CHAPTER 9

Material Requirement Planning

9.1 INTRODUCTION

Planning is a key managerial function. It provides the means to cope with a complex, dynamic and ever-changing environment. Planning is needed to bridge the gap between the present and the future. Planning provides the standards and guidelines needed for communicating, coordinating and controlling on going operations in an organisation. The planning process requires identification of priorities and an evaluation of the critical and limiting strategic needs. In this chapter we will discuss material requirement planning and MRP–II that is manufacture Resource planning.

9.2 DEFINITION AND NEED OF PLANNING

"Planning is deciding in advance what is to be done? When a manager plans, he projects a course of action for the future, attempting to achieve a consitent, coordinated structure of operations aimed at the desired results". The need for planning has become more compelling in the modern context because of the following factors:

(*i*) Rapid industrial and technological growth and the changing social trends.

(*ii*) Growing competition for both domestic and foreign business.

(*iii*) The rapid growth of computer-based information system.

(*iv*) Multiplicity of failure factors to be considered in judging the prospects and setting of goals of an organisation.

(*v*) Technological complexity of most products on the one hand, and rapid obsolescence on the other.

(*vi*) High risk and high investment requirements of modern business.

(*vii*) Developments of planning theory, techiniques and tools.

9.3 MATERIALS REQUIREMENT PLANNING

Materials Requirement Planning (MRP) is both an inventory control and a scheduling technique. It consists of a series of steps which start by determining what finished products are needed to meet the demand by time periods and are completed with a schedule of the finished product components needed at each assembly level for each time period.

Conventional Inventory Systems are based on:

1. ABC Classification
2. Reorder Models
3. Safety Stock
4. EOQ

In project situation and Batch production the demand is not uniform but discrete. As the demand is clearly known and so also the time frame, it is also known as "Time-phased requirements planning."

MRP is based on concepts like,

1. Independent versus Dependent demand
2. Lumpy demand
3. Lead times
4. Common use items

The use of computer has helped MRP to gain the popularity because the numerous calculations for the product explosion are readily handled with the direct access capability of the disc systems. The on-line capabilities of computers is another reason for the increased interest in MRP.

9.3.1 USES of MRP

MRP is the most useful scheduling technique for many industries engaged in fabricating and assembling products like automobiles, tractor-trailer equipments, Rail coaches, etc., It is especially suitable for situations where one or all of the following conditions exists:

(*a*) The final product is complex and made up of several levels of assemblies.

(*b*) The final product is expensive.

(*c*) The lead times for components and raw materials are relatively long.

(*d*) The manufacturing cycle is long for the finished product.

(*e*) Consolidation of requirements for several products is desirable so that economic lot sizes are applicable.

9.3.2 Objectives

The MRP seeks to achieve the following objectives in a simultaneous manner:

(*i*) Ensure the availability of materials, components and products for planned production and for customer delivery.

(*ii*) Maintain the lowest possible levels of Inventory.

(*iii*) Plan manufacturing activities, delivery schedules and purchasing activities.

9.3.3 MRP Systems

MRP systems are appropriate for dependent demand items. Dependent demand means the demand for an item is directly related to or the result of demand for "higher level" items. Thus, while the demand for the final product may be continuous and independent, the demand for lower-level, subordinate items composing the product tends to be discrete, derived and dependent. Dependent demand items need not be forecasted, but are calculated by the MRP system from the Master Schedule.

The key features of an MRP system are the time phasing requirements, generation of lower-level requirements, planned order releases and rescheduling capability. Planned order releases indicate when orders should be placed by purchasing and manufacturing activity. A flow diagram of how a typical MRP system may function is shown in Fig. 9.1.

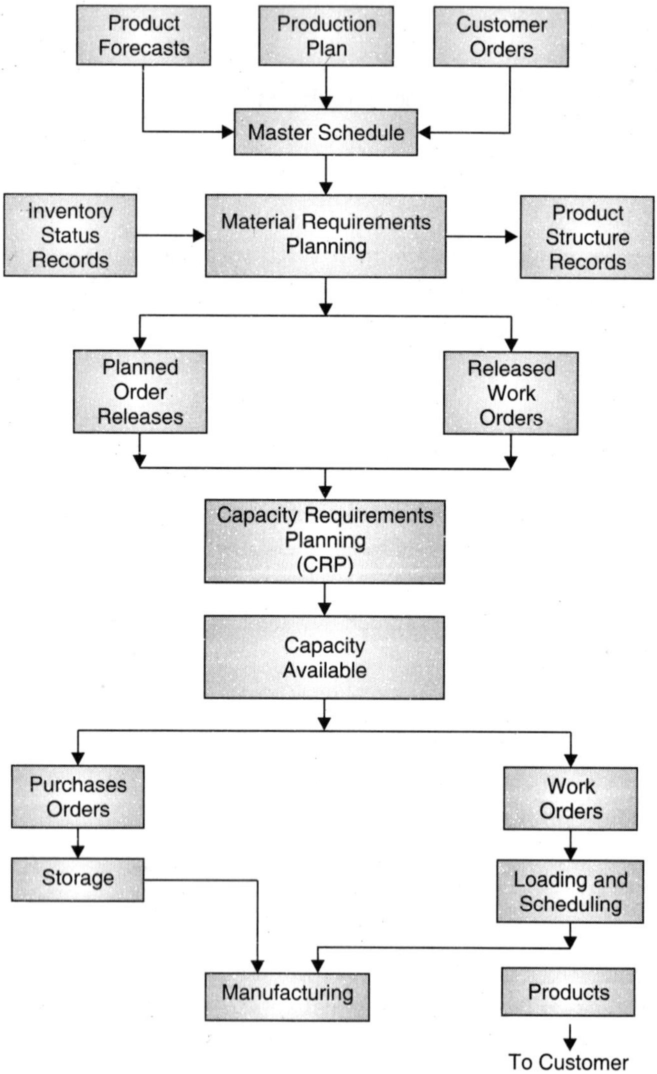

Fig. 9.1

MRP Inputs: The three major inputs of an MRP system are the master production schedule, the inventory status records and the product structure records.

The flow diagram of MRP diagram of MRP inputs is given in Fig. 9.2.

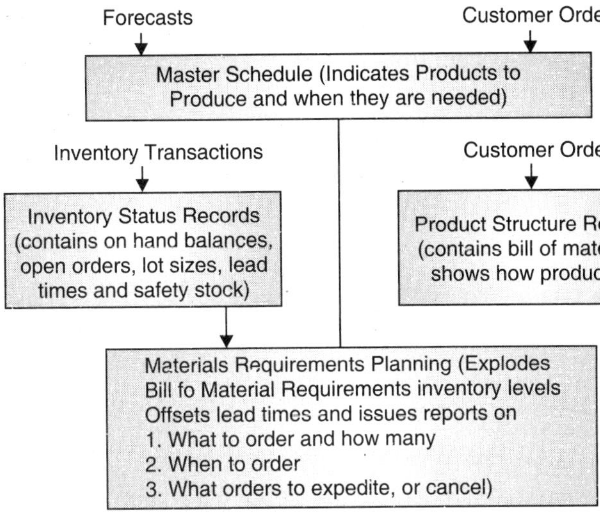

Fig. 9.2 MRP Inputs

MRP Outputs: The MRP system takes the master schedule for end items and determines the gross quantities of components required from the product structure records. Gross requirements are obtained by "exploding" the end item product structure record into its lower level requirements.

A schematic of MRP outputs is contained in Fig. 9.3.

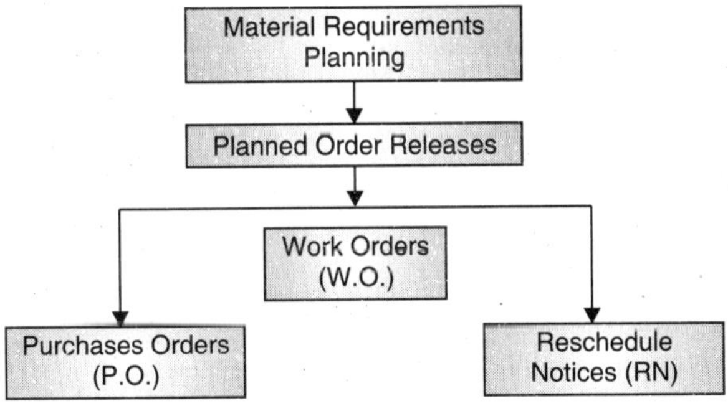

Fig. 9.3 MRP Outputs

Two basic purposes of planned orders are:

1. to generate the material requirement at the next lower level.
2. to project the capacity requirements.

Product Structure: MRP is well suited for fabrication and/or assembly type operations. The term "component" in MRP refers to all items below the product level, including sub assemblies, parts, raw materials, whether they are produced internally or obtained from suppliers.

A bill of material (BOM) is a list of all items, ingredients, or materials needed to produce an end item or product. It shows how much of what materials are needed in order to manufacture of product. It is through BOM that MRP successfully translates the master schedule into gross requirements below the end item level.

A schematic representation of a product structure is shown in Figure 9.4. The structure of product A defines the relationship among the various items that make up the product in terms of levels as well as parent/component relationships.

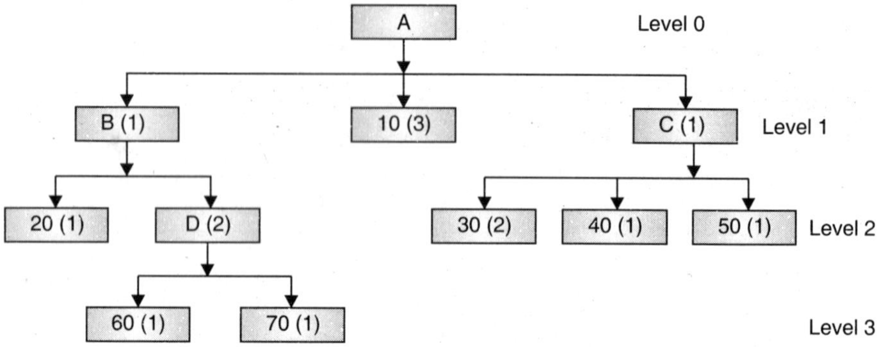

Fig. 9.4 Typical product structure.

The letters represent assemblies/sub-assemblies, and the numerical represent parts. The numbers in parentheses are the quantities required.

9.3.4 Material Requirement Planning Procedure

STEP 1 : Determine the Gross Requirement.
STEP 2 : Determine the Net Requirements.
Net Requirements = Gross requirements – Available Inventory.
STEP 3 : Develop the Master Schedule.
STEP 4 : Explode the Bill of Materials.
STEP 5 : Screen for ABC items.
STEP 6 : Determine net requirements for an item.
STEP 7 : Adjust requirements by Scrap or Shrinkage factor.
STEP 8 : Schedule planned orders.
STEP 9 : Explode the next level.
STEP 10 : Aggregate Demands and Determine Order Quantity.
STEP 11 : Write and place the planned orders.

9.4 MANUFACTURING RESOURCE PLANNING (MRP II)

This is a development that seeks to address some of the shortcomings of MRP. It includes all of the elements of MRP, it:

- is based around the Bill of Materials,
- uses a Master Production Schedule (MPS) as its starting point and
- uses the three steps of Explosion, Netting and Offsetting to create the initial schedule.

However MRP II includes the following four major developments from MRP:

1. Feedback: MRP II includes feedback from the shop floor on how the work has progressed, to all levels of the schedule so that the next run can be updated on a regular basis. For this reason it is sometimes called 'Closed Loop MRP'.

2. Resource Scheduling: There is a scheduling capability within the heart of the system that concentrates on the resources, *i.e.*, the plant and equipment required to convert the raw materials into finished goods. For this reason the initials 'MRP' now mean Manufacturing Resources Planning. The advantages of this development are that detailed plans can be put to the shop floor and can be reported on by operation, which offers much tighter control over the plant. Moreover loading by resource means that capacity is taken into account. The difficulty is that capacity is only considered after the MRP schedule has been prepared. It may turn out that insufficient time was allowed within the MRP schedule for the individual operations to be completed.

3. Batching Rules: Batching rules can be incorporated, indeed they have to be if resource scheduling is to take place. Most software packages offer a variety of batching rules. Three of the more important are 'Lot for Lot', 'EBQ' and 'Part Period Cover'.

- **'Lot for Lot'** means batches that match the orders. Therefore if a company is planning to make 10 of Product A followed by 20 of Product B, then the batches throughout the process will match this requirement. If both A and B require two of a certain sub assembly then that will be made in quantities of 20 of A and 40 of B. It is the batching implicitly followed in basic MRP.
- **'EBQ'** stands for Economic Batch Quantity. The batch size is calculated by a formula that minimises the cost through balancing the set up cost against the cost of stock.
- **'Part Period Cover'** means making batches whose size cover a fixed period of demand. A policy of making a weeks requirement in one batch is an example.

4. Software extension programmes: A number of other software programmes are included in the MRP II suite. Some of these are further designed to help the scheduling procedure. The most important is Rough Cut Capacity Planning (RCCP), an initial attempt to match the order load to the capacity available, by calculating (using a number of simplifying assumptions) the load per resource. Overloads are identified and orders can be moved to achieve a balance. This has been described as "knocking the mountains (the overloads) into the valleys (periods of underload)".

Other additions are designed to extend the application of the MRP II package. For example, it may include an option for entering and invoicing sales orders (Sales Order Processing). Another common extension is into stock recording and a third into cost accounting. A full MRP II implementation can therefore act as an integrated database for the company.

5. Data accuracy: This last development means that the company must put great emphasis on data accuracy. Errors in recording in one part of the system will result in problems for all the users. The suppliers of such systems encourage users to aim for accuracy of between 95% and 98%.

EXERCISES

1. Define planning. What is the need of planning?
2. Explain material requirement planning.
3. Explain manufacturing resource planning.

❑❑❑

CHAPTER 10

World Class Manufacturing

10.1 INTRODUCTION

In the face of increasing competition many companies are being forced to continuously improve their manufacturing operations. Today manufacturing strategy must support overall business strategy and indeed, many companies are looking to manufacturing to give them competitive advantage in the marketplace, example, the best quality product, the lowest product cost, the shortest lead-time, and the best customer service.World Class Manufacturing is a way of achieving this competitive advantage from manufacturing. It is the pursuit of superior performance in quality, lead-time, cost and customer service through continuous improvement using principles such as just in time manufacturing, total quality management, total productive maintenance, lean manufacturing. World Class Manufacturing is the best way to organize the company as to achieve the highest, possible level of organization of the business through the implementation from the organization of modern managing methods.

10.2 CONCEPT OF WORLD CLASS MANUFACTURING

World Class Manufacturing is a different set of concepts, principles, policies and techniques for managing and operating a manufacturing company. It is driven by the results achieved by the Japanese manufacturing resurgence following World War II, and adapts many of the ideas used by the Japanese in automotive, electronics and steel companies to gain a competitive edge. It primarily focuses on continual improvement in quality, cost, lead time, flexibility and customer service. World Class Manufacturing is a process-driven approach where implementations usually involve these philosophies and techniques; Quick Changeovers (SMED), Statistical Process Control (SPC), Just in Time (JIT), Visual Management, Streamlined Flow, Small lot sizes, Cross Functional Teams, Employee Empowerment, Part Families, Reducing Process Variation, Cellular Manufacturing, Zero Defects, and Total Preventive Maintenance (TPM).

Companies taking part in World Class Manufacturing strategies tend to focus on improving their operations, and to attempt to eliminate waste to create a lean business system. This approach results in higher productivity and better profits. However, these businesses will also focus on increasing the velocity of their total order volume to improve delivery without relying of excessive or increasing inventory levels. Sequential methods of performing work are being replaced with synchronized manufacturing methods to speed up the operations to compress time, and hierarchical departmentalized functions are being replaced by self directed work teams. The modern techniques of manufacturing like Just in time, tqm, tpm, six sigma, value analysis, lean manufacturing are discussed further in this chapter.

10.2.1 Advantages

(*i*) Reduction of time refitting,
(*ii*) cutting operating costs,
(*iii*) transparency and visualization approach,
(*iv*) continuous improvement in organizational performance,
(*v*) accelerate time-to product available,
(*vi*) meeting exceeding customer expectations,
(*vii*) outsourcing selected processes etc.

10.2.2 Limitations

(*i*) The implementation of world class techniques will often bring up resistance to change with management and employees. Resistance to change is often supported by the statement "we've always done it this way". The most resistance is often found in the lower and middle management levels. A process for change has to be implemented with the full participation of management and employees working together.

(*ii*) Capital investment is another huge issue for management teams. World Class Manufacturing has the approach of "use creativity before capital". However, sometimes improvements can be financially justified. This would be true, especially when improved methods require an organization to purchase new technology or equipment to achieve quick changeovers, more efficient cycle times, and give more flexibility in their operations. Executives will often choose to take the option to save on costs but this can be short sighted. They will become equally frustrated with the less than desirable outcome as a consequence of not financially supporting the improvements.

Like everything else, World Class Manufacturing is not a universal remedy or silver bullet, nor should it be. It is an operational approach that, if used correctly, it will provide new insights and allows a business to become more competitive in global markets. It is the ideal system to quickly design and develop new, higher quality products at lower costs, as well as delivering them with shorter lead times and better manufacturing processes.

10.3 TQM

It is a corporate business management philosophy which recognises that customer needs and business goals are inseparable. Following are some definitions of TQM:

1. TQM is a way of managing to improve the effectiveness, flexibility and competitiveness of a business.
2. TQM is both a philosophy and a set of guiding principles that represent the foundation of a continuously improving organisation.
3. TQM is an integrated approach in delighting both external and internal customers by meeting their expectations on a continuous basis through everyone involved in the organisational working on continuous improvement in all products, services and procedures along with proper problem solving methodology.

10.3.1 Characteristics of TQM

(*i*) Operation/process leading to end result.
(*ii*) Do your best as a team.
(*iii*) PDCA cycle—In built Deming Improvement cycle.
P – Plan D – Do C – check A – Act
(*iv*) Participative Management.
(*v*) Reward system should be based on need theory.
(*vi*) Suggestions system should be provided.
(*vii*) Non-financial incentives should be given for best suggestions.

10.3.2 TQM Models

I. Juran Trilogy

There are three phases of Juran Trilogy *viz*., Quality Planning, Quality Control and Quality Improvement:

(*a*) **Quality Planning:** It involves analysis of customer needs, product and process interpretation for implementing quality control.

Defects do not just happen, they are caused. There is a long list of causation patterns—Marginal product design, incomplete process engineering, poor vendor quality, lack of employee training, inadequate tool maintenance, employee carelessness, insufficient feedback of inspection data, sale of wrong applications etc.

Some defects are the result of a sudden sporadic flareup of a process that is normally well behaved. What has gone out of control, put it back in place.

Other defects are chronic in nature. They may be continued because there is permanent cause which we have not yet remedied or even discovered. Process needs to be improved.

(*b*) **Quality Control:** It involves operational techniques and activities aimed both at monitoring a process and at eliminating causes of unsatisfactory performance at relevant stages of quality loop in order to result in economic effectiveness.

The objective of quality control function at operators level, foremen level, inspectors and managers level is to involve with operational techniques and activities carried both at monitoring a process and at eliminating causes of unsatisfactory performance.

Thus quality control is concerned with the feedback of the comparative information in order to regulate the process. The entire process of quality control be readjusted before the outgoing product reaches to a limit of rejection.

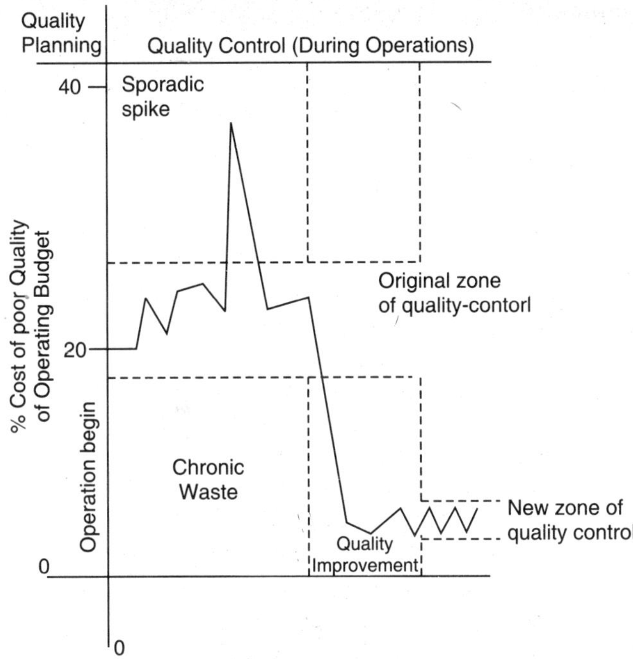

Fig. 10.1

(*c*) **Quality Improvement Approach:** It identifies sources of unacceptable variation and adjustments are made in product design, process procedures etc. there by eliminating sources of poor quality. The new zone of quality control translates to improve customer satisfaction, increased sales, lower unit material and labour cost, increase overall process capacity, a more motivated and secure workforce with greater self-esteem etc. As the improvement process progresses, lessons learned provide feedback to improve the planning process.

II. The Deming Chain Reaction

Fig. 10.2

III. The Quality Planning Road Map

Existing product and process ⟶ identify customer, discover customers needs ⟶ establish units of measure ⟶ develop product ⟶ optimize product design ⟶ develop process, optimize process capabilities ⟶ process ready to produce.

Fig. 10.3

IV. The Mckinsey TQM Model

Strategy → Top management should decide their strategy about quality

Structures → Restructuring is necessary

Systems → Should be developed

Staff → Should be utilized properly

Skills → Should be developed

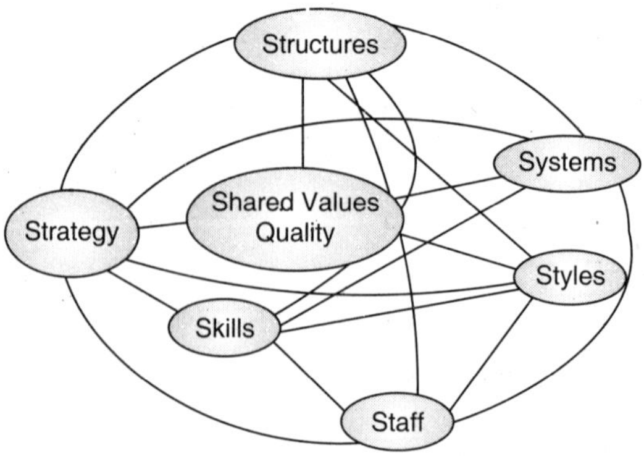

Fig. 10.4 Mckinsery 7-5 Framework.

Style → of functioning of the people, incharge of TQM will decide results. Shared Values → Should be developed.

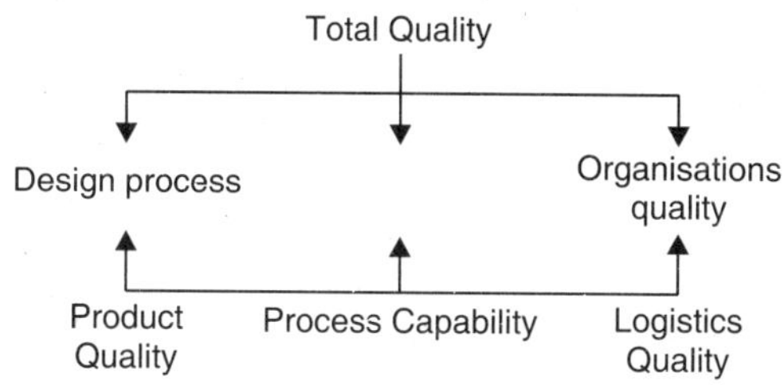

Fig. 10.5

V. The Crosby Program for TQM

(*a*) Management Commitment
(*b*) Quality improvement team
(*c*) Quality measurement
(*d*) Cost of quality evaluation
(*e*) Quality awareness
(*f*) Corrective action
(*g*) Establish an adhoc committee for the zero defects programme
(*h*) Supervisor training
(*i*) Zero defect day in a year/month/week
(*j*) Goal setting
(*k*) Error causes removed
(*l*) Recognition of people
(*m*) Quality councils
(*n*) Do it all over again.

VI. Anderson Consulting Model for TQM

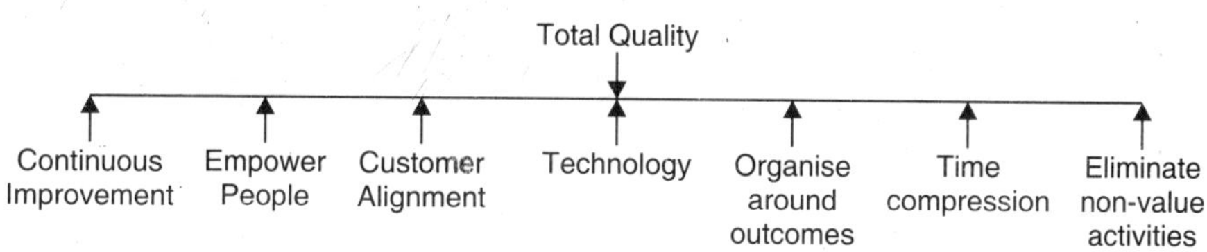

Fig. 10.6

VII. European Quality Award Model

VIII. Malcolm Baldrige Award

IX. Ambal Model for TQM

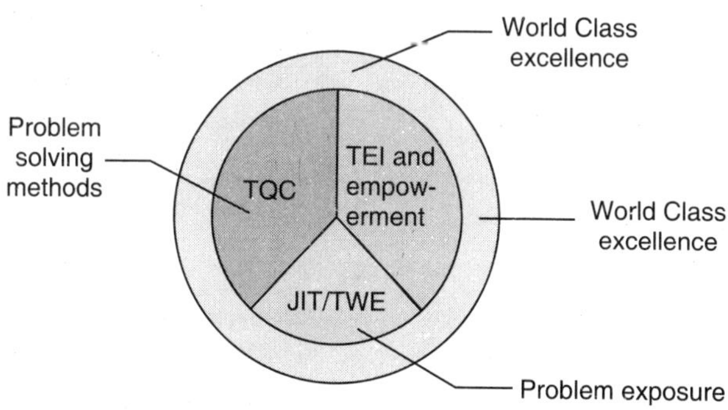

Fig. 10.7

JIT — To manage internal customers

TQC — For external customer's satisfaction

X. Integrated Model of TQM

Sohal, Tay and Wirth have discussed quality in terms of TQC rather than TQM but refer control to the management of quality at various stages of the process. There are five elements in this model:

- Customer (internal and external) focus
- Management commitment
- Total Participation: Total Employment Involvement (TEI)
- Statistical Quality Control
- Systematic problem solving process *i.e.,* Plan-Do-Check-Act (PDCA) cycle.

Fig. 10.8 The integrated model of TQM (Sohal, Tay and Wirth).

XI. OAKLAND Model for TQM

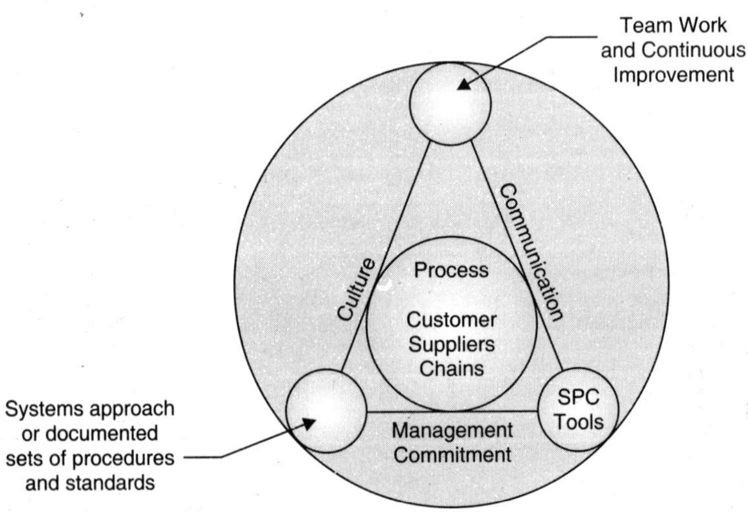

Fig. 10.9 The Oakland Model (Oakland, 1989).

Oakland has enumerated following thirteen stages to represent a gradual progression towards implementing a TQM based culture:

(*i*) Understanding quality (*ii*) Commitment to quality

(*iii*) Policy on quality (*iv*) Organisation for quality

(*v*) Measurement costs of quality
(*vi*) Planning for quality
(*vii*) Design for quality
(*viii*) System for quality
(*ix*) Capability for quality
(*x*) Control for quality
(*xi*) Team work of quality
(*xii*) Training for quality
(*xiii*) Implementation of TQM.

XII. The Building Blocks Model of TQM (Zairi *et al.*)

Zairi at the Management Centre of Bradford University has proposed three parts to discuss total quality.

The Foundation of Total Quality (TQ): Three elements of foundational block of TQM have been identified:

- Continuous Improvement
- Value addition at each stage of every operation
- Employee Involvement

The Pillars of Total Quality (TQ): These are the means by which the human creativity inputs can be converted into outputs which benefit the end customer. The pillars are represented by various quality systems which represent procedures, documentation, use of statistical process control (SPC), management control system, process flexibility provided by use of computers, workplace design, ergonomics and customer—(both internal and external)—supplier chain.

The Top of Total Quality (TQ): This top of building should shield the organisation from adverse external factors and protect it all the time. This top part should be weather proof (not affected by adverse changes in market place) and should not deteriorate (organisations are dynamic in nature and have to adapt to new market patterns). The activities of senior managers, therefore, should aim towards:

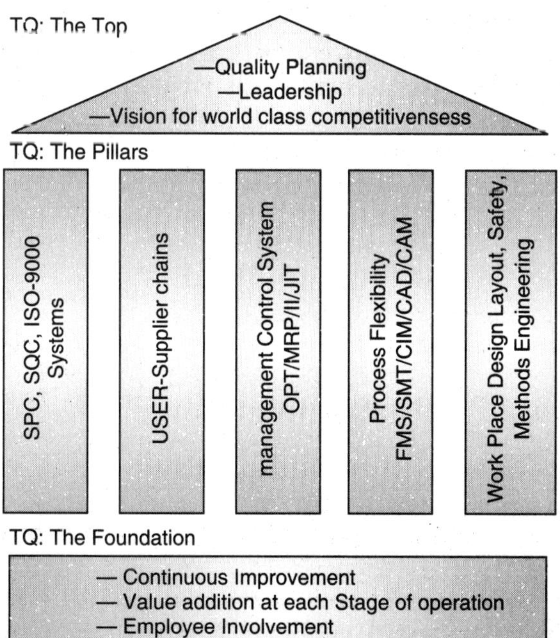

Fig. 10.10 The Building Blocks Model of TQM (Zairi, 1991).

- Quality planning
- Leadership
- Vision for World class competitiveness.

Figure 10.10. Depicts the TQM model as proposed by Zairi (1991):

TQM Drivers: Survival, strive for excellence, customers expectation, profitability, export drive, competition.

TQM System: Management Commitment, Quality tools and techniques, continuous education and training, customer satisfaction, recognition and rewards, teamwork, participation and involvement.

10.3.3 Advantages of TQM

TQM is advantageous for various reasons as mentioned below:

(*i*) Improved competitiveness.

(*ii*) Better profitability, better returns, reduction in operating cost.

(*iii*) Increased sales and market share.

(*iv*) Enhanced customer satisfaction.

(*v*) Improved cost effectiveness, reduced unit cost.

(*vi*) Reduction of rejection, scrap and wastage-reworking.

(*vii*) Better control on processes.

(*viii*) Successful new product launch.

(*ix*) Speedier new products introduction, more new models.

(*x*) Re-engineered process.

(*xi*) Improved quality. Continuous improvement is an important thing of TQM.

(*xii*) Improved productivity.

(*xiii*) Better relations; employees job satisfaction, employees participation, quality unites and inspires, status, recognition. Employer feel that he is committed to develop people.

(*xiv*) Lead time reduction.

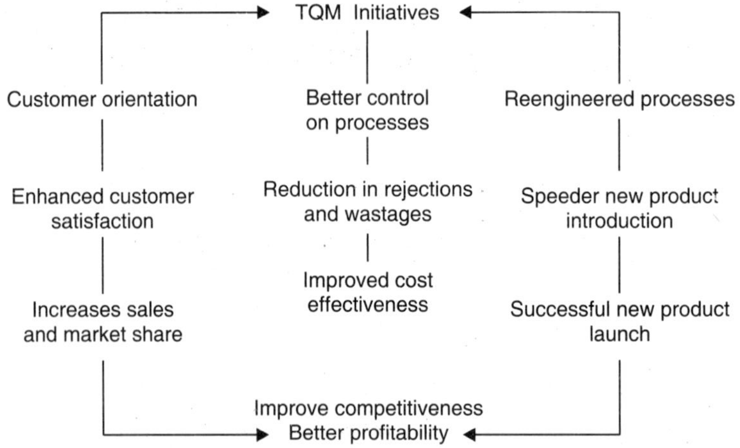

Fig. 10.11

10.3.4 Taguchi is 8-Point Approach

1. Identify the main function, side effects and failure modes.
2. Identify noise factors and the testing conditions for evaluating quality loss.
3. Identify the quality characteristic to be observed and the objective function to be optimized.
4. Identify the control factors and their alternate levels.
5. Design the matrix experiment and define the data analysis procedure.
6. Conduct the matrix experiment.
7. Analyse the data, determine optimum levels for the control factors, and predict performance under these levels.
8. Conduct the varification experiment and plan future actions.

10.3.5 Quality, Quality Assurance, Total Quality Control

Quality

(*i*) Quality is the fitness for the purpose. **—Dr. J.M. Juran**

(*ii*) Quality means getting every one to do what he has agreed to do and do it right the first time. **—Crosby**

(*iii*) Quality is the totality of features and characteristics of an entity or a product or service that bear in its ability to satisfy stated or implied needs. The implied needs are usually very vague, indicated or not indicated at all. Nontheless they are expected by the customer sometimes more than the standard: ISO 8402.

Quality Assurance

It means ensuring conformity of quality to the required basic specifications. All the planned and systematic activities implemented within the quality system and demonstrated as needed to provide adequate confidence that an entity will fulfill for quality requirements.

Total Quality Control

An effective system to integrate the quality development, quality maintenance and quality improvement efforts of various groups in an organization so as to enable production/service most economical for greater customer satisfaction.

10.3.6 Kaizen

Kaizen means improvement and an ongoing improvement which involve everyone in the organization—top, middle and lower managers, supervisors and workers. Kaizen is a Japanese word which means continuous improvement. Kaizen strategy is the key to Japanese competitive success. Kaizen is the most commonly used term in Japanese newspapers, TV, radio and other media modes. It is very frequently used there by goverment officials, politicians, traders, managers and diplomats. There is a strong belief among companies that they should devote at least 50 percent of their attention to Kaizen. Kaizen has its emphasis on process oriented way of thinking and management system that supports and acknowledges peoples' process-oriented efforts for improvement.

Kaizen practitioners view quality as an endless journey, not a final destination. They are always experimenting, measuring, adjusting and improving. Rather than assuming that zero defects means perfection, they search for potential and actual trouble sports for improvement of product design and process.

There are four general avenues for continuous improvement for any organization, These include:

1. Improved and more consistent product and service quality.
2. Faster cycle time (cycles ranging from product development to order processing to payroll processing).
3. Greater flexibility (for example, faster response to changing customer demands and new technology).
4. Lower costs and less waste (for example, eliminating needless steps, scrap, rework and non-value adding activities).

10.4 TOTAL PRODUCTIVE MAINTENANCE

Total productive maintenance (TPM) is a set of techniques that are aimed at ensuring that all machines in a production process are always available when they are needed and never break down on the job. This is a neat trick, especially in factories where shifts work around the clock.

A key principles of TPM is to elevate the humble maintenance task to be a serious part of the production process, rather than a bunch of fix—it mechanics or a 'when there is time' activity.

TPM is a manufacturing led initative that emphasises the importance of people. It is a 'continuous improvement' philosophy where production and maintenance staff work together.

TPM brings maintenance into focus as a neccessary and vital part of the business. It is no longer regarded as a non-profit activity. Down time for maintenance is scheduled as a part of the manufacturing day and, in some cases, as an integral part of the manufucturing process. The goal is to hold emergency and unscheduled maintenance 'to a' minimum.

10.4.1 Need of TPM

TPM was introduced to achieve the following objectives. Some of them are listed below:

1. Aviod wastage in a quickly changing economic environment.
2. Producing goods without reducing product quality.
3. Reduce cost.
4. Produce a low batch quantity at the earliest possible time.
5. Goods send to the customers must be non-defective.

10.4.2 TPM-History

TPM is a innovative Japanese concept. The origin of TPM can be traced back to 1951 when preventive maintenance was introduced in Japan. However the concept of preventive maintenance was taken from U.S.A. Nippondenso was the first company to introduce plant wide preventive maintenance in 1960. Perventive maintenance is the concept wherein, operators produced goods using machines and the maintenance group was dedicated with work of maintaining those machines, however with the automation of Nippondenso, maintenance became a problem as more maintenance personnel were required. So the management decided that the routine maintenance of equipment would be carried out by the operators. Maintenance group took up only essential maintenance works. Thus Nippondenso which already followed preventive maintenance also added autonomous maintenance done by production operators. The maintenance crew went in the equipment modification for improving reliability. The modifications were made or incorporated in new equipment. This lead to maintenance prevention. Thus preventive maintenance along with maintenance prevention and maintainability improvement gave birth to productive maintenance. The aim of productive maintenance was to maximise plant and equipment effectiveness to achieve optimum life cycle cost of production equipment.

By then Nippondenso had made quality circles, involving the employees participation. Thus all employees took part in implementing productive maintenance. Based on these developments Nippondenso was awarded the distinguished plant prize for developing and implementing TPM, by the Japanese Institute of plant Engineess (JIPE). Thus Nippondenso of the Toyota group became the first company to obtain the TPM certification.

10.4.3 TQM Vs TPM

The TPM program closely resembles the popular Total Quality Management (TQM) program. Many of the tools such as employee empowerment, benchmarking, documentation etc. used in TQM are used to implement and optimize TPM. Total commitment to the program by upper level management is required in both programmes. Employees must be empowered to initiate corrective action, and a long range outlook must be accepted as TPM may take a year or more to implement and is an on-going process. Changes in employee mind set toward their job responsibilities must take place as well. The difference bteween TQM and TPM is:

Categary	*TQM*	*TPM*
Object	Quality (output and effects)	Equipment (Input and cause)
Target	Quality for product and process	Elimination of losses and wastes.
Means of attaining goal	Systematize the management. It is software oriented.	Employees participation and it is hardware oriented.

10.5 VALUE ANALYSIS

Value analysis discovers new tactical alternatives. It is a special type of cost reduction technique which unlike other methods does not bindly accept the product design. Rather, it critically investigates and analysis the different aspects of materials, design and production of each and every component of the product. A component, to be a part of the product must perform some function and serve some purpose. Value analysis examins the design, function and cost of each and every component in order to produce it economically without decreasing its utility, function or reliability.Value analysis developed as a cost reduction technique in U.S.A. in 1947. The credit for it goes to Lorly D. Miles who was working at G.E. and who, subsequently became the president of SAVE (Society of American Value Engineers) also. Value engineering got developed because of the inherent desire in the man to make cheaper and to sell cheaper; keeping the utility of the products same.

10.5.1 Value

The term 'Value' is often confused with the term 'Cost' or 'Price'. But in reality value differs from both price and cost in the sense that it is the cost proportionate to the function. This can be written as: Value or worth = Function or utility/cost.

If we want to increase the value, either we have to increase the utility of the product with the same cost or by decreasing the cost for the same function. Value can be divided into the following categories:

1. *Esteem Value:* It is the desire to secure or own a certain article for status or prestige.
2. *Exchange Value:* It is the value that makes the object possible of being traded for other items.
3. *Use value:* It is property, which accomplish a use, as a work, or a service.
4. *Cost Value:* It is sum of all the costs involved in manufacturing the product, *i.e.*, material, labour, overheads etc.

10.5.2 Objectives of Value Analysis

The major objectives of value analysis are:

1. Saving in product cost.
2. Increase the utility of the product.
3. Modify and improve product design.
4. Quality promotion.
5. Elimination of unnecessary processes
6. Increase the profits.
7. Simplify the product.
8. Use cheaper and better materials.
9. Improving company image.

10.5.3 Value Analysis Procedure

The basic steps in conducting value analysis are:

1. Identify the product for which value analysis is to be carried out. The product should be one which if redesigned can add to the sales income and would not become obsolete in future. Value analysis may be applied to the product as a whole or to only some of its component parts.
2. The information relevant to the product like drawing and technical specifications, manufacturing process, bill of materials, detailed cost break

up, performance report, quality, procurement including whether produced in your undertaking or purchased from outside and production problems are to be collected.

3. Identify the primary, secondary functions of the product. Specify value content of each function and identify the high cost areas.
4. Generate ideas and create all the possible alternatives by application of brain storming and other creativity techniques. All suggestions feasible or non-feasible are recorded without any criticism rather persons are encouraged to express their views freely.
5. The alternatives developed are analysed. The ideas technically sound and involving lower cost are further developed.
6. Detailed development plans are made for those ideas which during step 5 appeal most suitable and promising. Development plans comprise of drawing the sketches, building of models, and conducting discussions with the purchase section, finance section, marketing division etc.
7. The best found idea under step 6 is converted into a prototype manufacture which ultimately goes into operation and its result are recorded.

10.5.4 Advantages of Value Analysis

1. Value analysis reduces production cost and adds to sales income of the product.
2. Value analysis is a much faster cost reduction technique.
3. It requires little expenditures because value analysis team can be formulated out of the staff available in different sections like purchase, production, finance etc.

10.6 JUST-IN-TIME

Just-in-time is an approach to minimise waste in manufacturing in the form of time, materials, energy and errors. There are many definitions of JIT but something which is common to all these definitions is that it is a broad philosophy of pursuing zero inventories, zero transactions and zero disturbances, *i.e.*, execution of the schedule day-in-day out. The major issue in any JIT system is maintaining flow times, work must flow through the factory quickly so that detailed tracking is not required. The flow of materials through the factory depends upon the operation cycle time, set-up time, queue time, handling system, layout, line balancing, machine maintenance and other such factors. The worker's attitude is very vital as they can remove these road blocks to increase speed and reduce cycle time. JIT displays an excellent ability to adapt to the changes in demand and help reduction of work in process inventory.

10.6.1 Objective of JIT

The objective of the JIT are as follows:

(*i*) To achieve zero defect goal in manufacturing.

(*ii*) To reduce the set up times and lot sizes.

(*iii*) To focus on continuous improvement.

(*iv*) To concentrate on involving workers and using their knowledge to a greater extent.

(*v*) To layout equipment in such a way so as to minimise both travel distance and inventories between the machines.

(*vi*) To-reduce inventories and thus economise on inventory carrying costs.

(*vii*) To eliminate waste such as long set up time, zig-zag material flow, scrap, machine breakdown, higher stocks, rework, inspection etc. JIT aims at identifying any problem related to waste and solve that through total employees involvement.

(*viii*) To crosstrain the workers in multi-functions to maintain and enable them to run several machines at a time.

(*ix*) To eliminate all non-value adding activities by systematically identifying these.

10.6.2 Benefits of JIT

1. The lot sizes are cut to less than economic order quantity in JIT manufacturing.
2. The machine set-up cost and time are cut.
3. Scrap reduction and quality improvement effects are maximum.
4. Customers get better quality goods at least costs.
5. The time and money spent on rework drop.
6. The workers get motivated to improve.
7. There is no particular need of supervisors.
8. The need to avoid errors is apparent under JIT system, which improves worker's feelings of responsibility.

10.6.3 Elements of a JIT System

Before going into details of the JIT system, we shall briefly summarise how the JIT system works as a whole. This will be done by starting with the various JIT elements such as master schedule and working backwards through the production proccess to the vendors. The following text describes the major elements of a JIT system.

Master Schedule

In JIT the master schedule (or final assembly schedule) is planned 1 to 3 months in advance to allow work centres and vendors to plan their respective work schedules. Without the current month the master schedule is levelled on a daily basis. In other words, the same quantity of each product is produced each day for the entire month. Furthermore, small lots (Preferably equal lot sizes) are scheduled in the master schedule to provide a uniform load on the plant and vendors during each day. The advantage of this kind of master scheduling is that it provides nearly constant demands on all downstream work centres and vendors.

Kanban

JIT uses a simple parts withdrawal system (a Japanese term called *Kanban*) to pull parts from one work centre to the next. Parts in the workshops are kept in small containers. Only a specific number of these containers are provided to the workers in each workplace. When all the containers are filled, the machines are shut off. No more parts are produced until the subsequent (using) work centre provides another empty container. Thus, work-in-progress inventory is limited to available containers. The parts are provided only when these are needed. The final assembly schedule, pulls parts from one work centre to the next just-in-time to support production needs. If a process stops because of machine breakdown or quality problems, all preceding processes will automatically stop. Alternatively, they will stop when their part's containers become full. The *kanban* system is explained in Fig. 10.12 and 10.13.

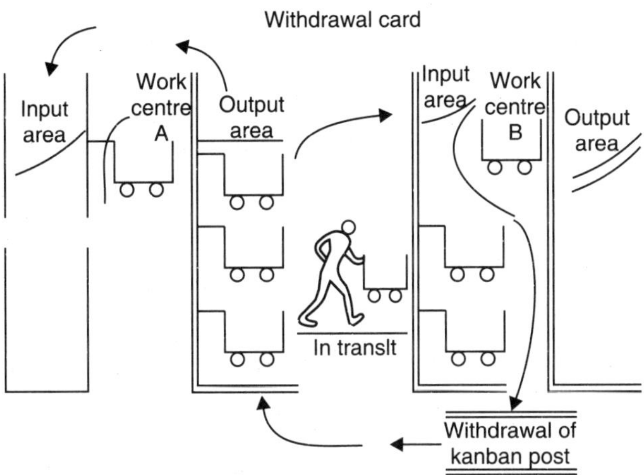

Fig. 10.12 Kanban System.

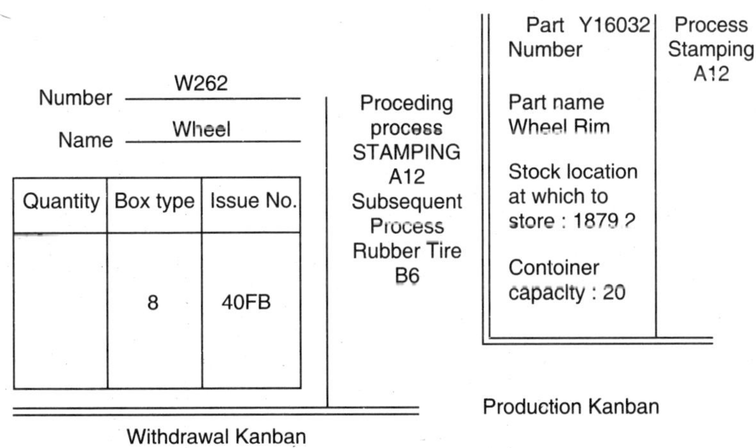

Fig. 10.13 Kanban Cards.

Set-up Time

The objective of JIT is to produce parts in a lot size of 1. In many cases, this is not economically feasible because of the costs of set-up compared to inventory carrying costs. The set-up time is not taken as given but rather considerded a cause of excess inventory. Low set-up time results in small, economical lot sizes and shorter production lead time. Down the set-up time for machine is—the key to the JIT system. With shorter lead times and less material in process, the production system is also much more flexible to changes in the master schedule.

Multiple Function Workers

With emphasis on quick changeovers and smaller lots, the multi-functional worker is required. Cross-training is needed under the JIT system so that a worker can switch from one machine to the next. Besides the workers can carry out their own as well as the other's jobs with respect to set-up time and maintenance. This requires a broad range of skills than the traditional manufacturing. JIT requires not only broader skills but also much greater teamwork and coordination since inventory is not available to cover up problem in the system. The entire production system must be more closely coordinated by the workers.

Layout of Plant

The layout of the plant is much different with JIT since inventory is held on the shopfloor and not put in a store room between the processes. Inventory is kept out in the open, so that it is readily available to the next process. Since inventory is typically kept low (only a few hours), the supply-plant can be kept much smaller because of the reduced storage space needed. A research study showed that only one-third of the space was needed compared to that in conventional plants (See Fig. 10.14).

Emphasis on Quality

Quality is absolutely essential with a JIT system. Not only that defects produce waste, but they can also grind the production process to a halt. Since there is no inventory to cover up for mistakes, perfect quality is required by a JIT system. JIT, however, facilitates very good quality since defects are quickly discovered by the next process. Quality problems rapidly gain plantwide attention as the production line stops when problems occur. A JIT system is designed to expose errors and get them corrected rather than covering them up with inventory.

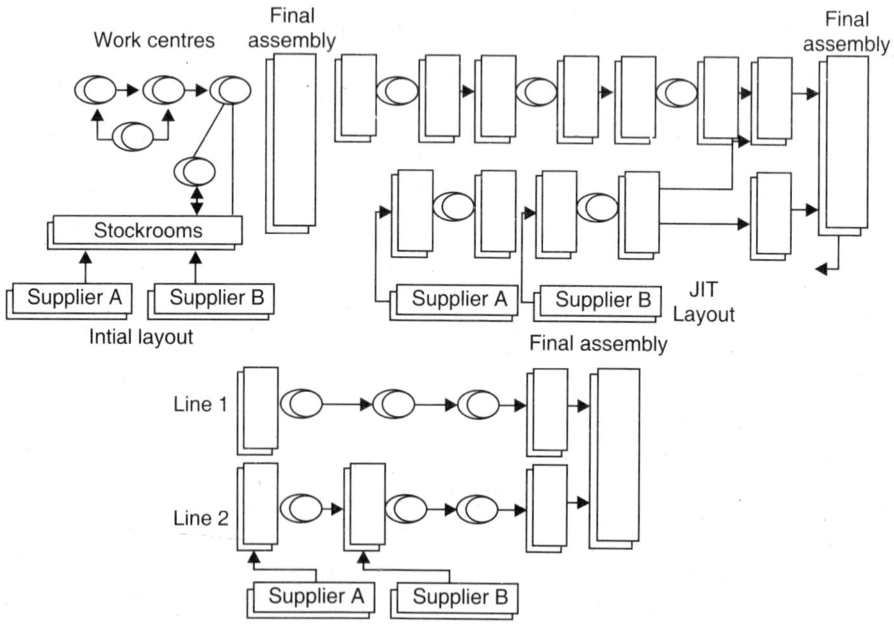

Fig. 10.14 Plant Layout under JIT.

Vendors' Relations

Finally, vendor's relations are radically changed by a JIT system. Vendors are asked to make frequent deliveries (as many as four times per day) directly to the line. Since vendors are viewed as an extension of the plant, they receive *Kanban* containers just as in-plant work centres do. Changes in shipping procedures and close proximity of vendors are often required to integrate them effectively with JIT procedures. Vendors are also required to deliver perfect quality. A revolution is required in the way that we usually think of our vendors. We need to think of them as partners rather than adversaries.

10.6.4 Kanban System

Kanban is a method of production authorisation and material movement in the JIT system. In the Japanese language, *Kanban* means a marker (card, sign, plaque, or other device) used to control the sequencing of jobs through a sequential process. *Kanban* is a sub-system of JIT. The two terms are not synonymous as some authors have indicated.

The purpose of the *Kanban* is to signal the need for more parts and to ensure that these parts are produced in time to support subsequent fabrication or assembly. This is done by pulling parts through from the final assembly line.

Only the final assembly line receives a schedule from the dispatching office. This schedule is nearly the same from day-to-day. All other machine operators and vendors receive production orders (*Kanban* cards) from the subsequent (using) work centre. If production stop for a time in the using work centres, the supplying work centres will also stop since they will no longer receive *Kanban* orders for more material.

The *Kanban* system is a physical control system consisting of cards and containers. Let us assume for the purpose of discussion that eight containers are used between work centres A and B (A supplies to B), and each container holds exactly 20 parts. The maximum inventory which can exist between these two work centres is then 160 units (8 × 20) since production at work centre A will stop when all containers are filled.

In the normal course of events, the eight containers might be distributed as shown in Fig. 10.12. Three containers are located at work centre A in the output area filled with parts. One container is presently being filled at work centre A by the machine. One full container is being moved from A to B, two full containers are sitting in the input area of work centre B, and one container is being used at B. These eight containers are needed since work centre A also produces parts for other work centres, machines at A may breakdown, and move times from A to B are not always exactly predicatable.

To control movement of the containers, there are two main types of *Kanban* cards-production cards, and withdrawal (move) cards. These cards are used to authorise production and to identify parts in any container, *Kanban* cards may be made of paper, metal, or plastic.

Here is how the Kanban system works, assuming containers are moved one at a time. When a container of parts is emptied at work centre B, the empty container and associated withdrawal card are taken back to work centre A. The production card from a full container of parts is removed and replaced by the withdrawal card. The production card is then placed in the *Kanban* receiving post at work centre A, thereby authorising production of another container of parts, and it is the empty container which is left at work centre A.

The full container of parts and its withdrawal cards are moved to work centre B and placed in the input area. When this container of parts is eventually used, its withdrawal card and the empty containers are taken back to centre A and the cycle is repeated.

All parts are neatly placed in containers of a fixed size. As empty containers accumulate, it is clear that the producing work centre is logging behind. On the other hand, when all containers are filled, production is stopped. The production lot size is exactly equal to one container of parts.

The number of containers needed to operate a work centre is a function of the demand rate, container size, and the circulation time for container. This is illustrated by the following formula:

$$\text{No. of containers } n = \frac{\text{DT}}{\text{C}}$$

where n = total number of containers

D = demand rate of the using work centre

C = container size in number of parts, usually less than 10 percent of the daily demand.

T = time for a container to complete an entire circuit: filled, wait, moved, used, and returned to be filled again. This is also called lead time for work centre A.

Suppose that demand at the next work centre is 2 parts per minute and a standard container holds 25 parts. Also assume that it takes 100 minutes for a container to make a complete circuit from work centre A to work centre B and back to A again including all set-up, run, move, and wait time. The number of containers needed in this case is 8.

$$n = \frac{2 \times 100}{25} = 8$$

The maximum inventory is equal to the container size × number of containers

$$= 8 \times 25 = 200 \text{ units.}$$

$$\boxed{\text{Maximum inventory} = n\text{C} = \text{DT}}$$

The maximum inventory is equal to the container size times the number of containers used. This is done by reducing the time, required to circulate a container including its machine set-up time, run time, wait time, or move time. When any of these times has been reduced, then management can remove *Kanban* cards from the system and a corresponding number of containers. It is the responsibility of managers and workers in a JIT system to reduce inventory by a constant cycle of improvement. Reducing lead time is the key. *Kanban* links all work centres in a production facility and the vendors. All material is pulled through by the final assembly schedule, based on a highly visible shopfloor and vendor control system.

10.7 SIX SIGMA

Six sigma is a business concept that answers customers demand for high quality and defect-free business processes. Customer satisfaction and its improvement should be the highest priorities of any business.

Six sigma was born when Motorola published its six sigma quality programme in 1987. Six sigma was developed by Mikel J. Harey. The programme gained publicity when Motorola won the Malcolm Baldrige quality award.

Sigma (σ) is a character of the Greek alphabet which is used in mathematical statistics to define standard deviation. The standard deviation indicates how tightly all the various examples are clustered around the mean in a set of data.

Six sigma is a business method of improving quality by removing defects and their causes in a business process activities. This method uses various statistical tools to measure business processes. "Six sigma means that there are 3.4 defects out of every 1 millon units of output." Six sigma is telling companies in clear, accurate mathematical terms how good—or more likely bad their quality levels are, how much they can improve, and what progress they are making on that journey. Six sigma strategists are leveraging this knowledge to consummate exciting improvements in quality—not just on shop-floor, but all over their organizations.

Six sigma is carried out as projects. Most common type is the DMAIC (Define, Measure, Analyze, Improve, Control) method. First, the project and process to be improved are defined after which the performance of the process is measured. The data is then analyzed and bottlenecks and problems identified. After analysis, improvement programme is defined and defects removed. This development programme is controlled by management. After DMAIC Circle, it is time to define a new project.

10.7.1 Six Sigma Implementation

It is reasonable to carry out the six sigma implementation as a six-phase process. Phases from one to four are to establish six sigma, and phases from five to six are for realization of six sigma. Six sigma implementation model is given in Fig. 10.14.

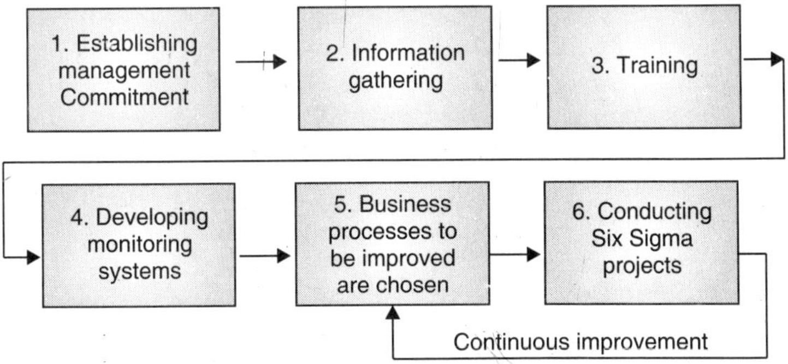

Fig. 10.14 Six sigma implementation model.

First phase is to estabish commitment of supreme command. This should be followed by development of a management infrastructure to support the Six Sigma. One tool monitor the change in management is to use QPR Score Card.

Second phase is information gathering. This translates into intensity communication with customers. Suppliers, and employees. Information about the conditions of the processes which should be improved. For example, supply chain to obtained from customers and suppliers. Analysis of the information helps to idently the obstacles standing in the way of success.

Training the whole staff in the oraganization is the *third phase* on the way to six Sigma. The training needs are assessed, and the training is conducted from top-to-bottom. Training levels can be, for example, black belt and green belt. *Black belts* are the all-day problem solvers who also operate as team leaders in Six Sigma projects. *Green belts* are the team members in Six Sigma projects.

Fourth phase of the Six Sigma implementation process is to develop a monitoring system. Because the main *goal* is continuous improvement, the framework of the management has to support that end. After the framework is in order, it is time to develop a monitoring system. Adequate measures should be created for strategic goals and key business processes. Measures can be internal *and/or* external, for example, customer satisfaction. One framework element is, for example, a balance score card.

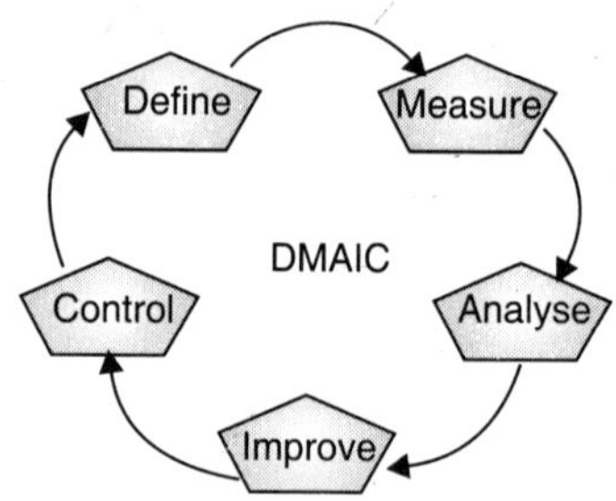

Fig. 10.15 DMAIC Process.

The realization of the Six Sigma starts from *phase five* when business processes to be improved are chosen. The current key process in general level are mapped and problems identified, and valueless activities and sub-processes are terminated. Then the processes that need improvement are chosen and the Six Sigma project is defined. One project concept is DMAIC where the problems of the process that need improvement are first defined and then the goals that should be achieved during the project are determined. The existing system needs to be measured. Establish valid and reliable measures and carry out a research. Analyse the findings and define the gap between "as-is" and "should-be".

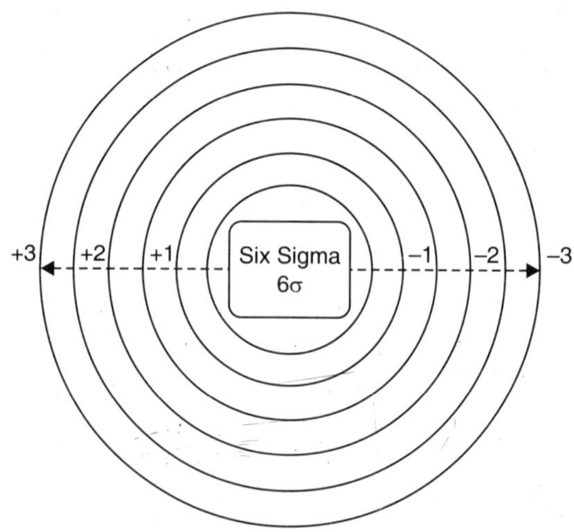

Fig. 10.16 The Enigma of Six Sigma.

After analysis, the system starts to create new ways of doing things. Improve the processes and validate them by simulation and statistical methods. Before implementing the imporvement create a project plan and communicate the changes through the organization. Institutionalize the new imporved system by modifying the management system. This also mean's modifying compensation and incentive system, Enterprise Resource Planning (ERP) and budgets. Establish a proper documentation system, for example, ISO 9001: 2000, and monitor the system. Finally, transfer the information. Which means that findings of the project are assessed for relevence to other projects.

Sigma	Defect Rate (PPM)	Cost of poor quality (% of sales)	Competitive level
6σ	3.4	<10%	World Class
5σ	233	10%–15%	World Class
4σ	6,210	15%–20%	High Quality
3σ	66,807	20%–30%	Industry Average
2σ	3,08,537	30%–40%	Indurstry Average
1σ	6,90,000	>40%	Non-competitive

Fig. 10.17 The Sigma Scale of Defects.

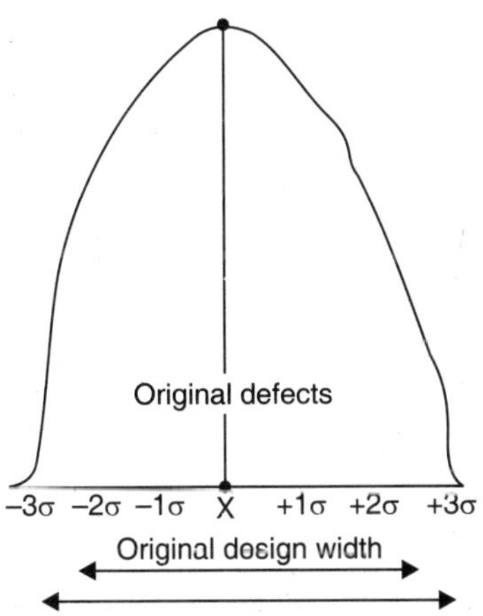

Fig. 10.18 The Mathematics of Six Sigma.

Most of the output of a process will meet your specification (call it X). But some will deviate, to varying extents, measured by the standard deviation(s). So some units will have a specification of X ± Is; some X ± 2s; and some X ± 3s.

10.7.2 Roles in Six Sigma

Six Sigma is a team job rather than an individual job. Thus different roles are required to be assigned to different people. Table 10.1 summarizes the roles that must be followed in general.

Table 10.1

Title	*Role*	*Experience and skills*
Black Belt	It is the most critical role. Without a qualified black belt, the procces may fail. People are skilled in Six Sigma methodology, tools and processes. This belt serves generally from 18 months to two years, and carries out 8 projects.	This team is usually developed from within the organization, but is re-assigned full time to, six sigma. Usually a middle manager is developed from this belt. The incumbent must be an expert in problem solving, data collection and analysis. He must be organizational savvy. The focus is on leadership/coaching, and understanding administrative systems and processes.
Master Black Belt	This team is Coach., mentor, and trainer to Black Belts People are expert in sit Sigma methodology, tools and processes. This belt offers hands-on help to Black Belts.	Master black belt person is generally a professional six sigma consultant. He must process substantial experience with all the areas mentioned under Black Belt..
Green Belt	This belt is similar to Black Belt, but holds a regular position while participating in or leading a Six Sigma team.	Similar to Black Belt
Champion or Sponsor	Champion initiates and sponsors a Black belt or team project. This belt is directly accountable for project results. They align project with business goals.	Champion or Sponsor must be Executive-level or key operational manager. He is often member of organization's Leadership council or steering committee.
Implemenatation Leader	This belt orchestrates entire Six Sigma effort within an organization.	Implementation leader is a corporate VP reporting to Chief Executive. He is, often a seasoned professional in quality or organizational improvement hired specifically for Six Sigma effort.

10.8 5S

5S are techniques to improve the workplace for better and efficient workplace working conditions. Thus 5S results in reduced wastages, less accidents, better quality of product and hence increased profits. These techniques were developed in Japan and are set of 5–words in Japanese (each word representing one principle with each word starting with '*S*'. These words are:

1. **'SEIRI' (Sorting):** It means to distinguish the necessary and unnecessary in accordance with specific rules and principles.
2. **'SEITON' (Make Convenient):** It arrange the necessary items for easy and convenient use. A place for every thing and everything in the place is the basic philosophy of 2S.

3. **'SEISO' (Clean Up):** It is a philosophy and commitment to be responsible for all aspects of the things you see and to ensure that they are in the tip-top condition.
4. **'SHIEKESTSU' (Standardisation):** Standardization involves keeping things organised neat and clean even in personal, pollution related aspects etc. Prepare manuals for standard cleaning procedures, work habits etc.
5. **'SHITSUKE' (Promoto good habits):** It is discipline. Discipline means to get accustomed to the operations and other in compliance with standards.

10.9 LEAN MANFURING

Lean is the term used to describe the production systems and methods developed by the Toyota Company in the post war years. Lean comes about by focusing on the reduction of the non-valued activities within the organization. Lean is *all* about achieving more value by applying fewer resources more effectively and efficiently through the continuous elimination of non-valued added activities *or* waste.

With its origins in the automotive industry the concept of Lean is widely applied in manufacturing industry but is not restricted solely to manufacturing activities and instead can be easily applied to the whole organization's administrative functions as well as to the service sector too. Lean can be and often is extended beyond the organization's boundary to encompass the supply chain.

10.9.1 Principals of Lean Manufacturing

1. **Identify and understand what creates value for the customer:** By accepting that only a fraction of the total time and effort expended in the organization actually adds value for the customer; understanding what the customer wants is the first step towards identifying all the non-value added activities or waste within the organization.
2. **Identify, understand and map all the process steps required to fulfill the customer:** By documenting the entire set of activities across all parts of the organization involved in delivering products or services to the customer allows for an understanding how value is delivered to the customer in the current state makes it possible to identify the wasteful activities that can then be targeted for elimination.
3. **Make the value-added actions within processes flow:** Flow is about creating value for the customer with the minimum number of process steps with the minimum number of delays between those process steps and the minimum number diversions along the way. The value stream is used to identify wasteful activities to be eliminated; flow concerns the rethinking or reorganization of the remaining activities to further eliminate waste.
4. **Only make what the customer wants or Pull:** If one objective for creating flow within processes is to minimize their throughput time. A clear second is to ensure that processes are only initiated when customer demand requires it. Essentially, this may mean only making products or delivering services at the point the customer demands; just-in-time. As such as many process steps in

the value stream should be triggered when there is a firm customer requirement.

5. **Pursue Perfection through Continuous Improvement:** Creating flow and pull begins with often radical changes to individual process steps, but the gains delivered by Lean principles really become significant as they are applied with increasing scrutiny. As this happens more and more layers of waste become visible and the cycle continues towards a theoretical end point of perfection; where every asset and every action adds value for the customer. Through continuous improvement the organization should strive for incremental improvements in the value stream every single day.

In applying the five principles of Lean your organization will encourage a culture of continuous improvement; it will become "the way things are done" instead of "that's the way we've always done it". You will be ensuring that every asset and activity are driving towards the overall organizational strategy through constant review of processes and the targeted elimination of waste to ensure that they constantly and consistently deliver value to the customer. Lean allows the organization to maintain high levels of customer service whilst being able to grow and flex with a changing environment and it does this through implementing sustainable change.

10.9.2 Benefits of Lean Manufacturing

Lean essentially aims to compress time required to deliver value to the customer.

Typically, if you quarter lead-times, you will double productivity and reduce costs by 20%.

This is known as the ¼:2:20 rule. Typical results are as follows:

- Halving of lead-times; doubling of stock turnover.
- 30% to 50% reduction in floor space requirements.
- 20% to 40% increase in overall equipment effectiveness.
- 20% to 25% labor productivity gains.
- Reduction in administration and co-ordination roles.

EXERCISES

1. Explain world class manufacturing concept.
2. Define TQM.
3. Explain total productive maintenance.
4. Differentiate TQM and TPM.
5. Define KAIZAN.
6. Explain Kanban System of JIT.
7. What are '5S' techniques? Explain these techniques.
8. Explain lean manufacturing.

❑❑❑

Question Bank

IMPORTANT QUESTIONS
(*2 Marks*)

1. Define production and operation management.
2. Differentiate product and service.
3. What do you mean by 'Routing'?
4. What is Job type production system?
5. What is aggregate production planning?
6. Define total quality management.
7. What is meant by value analysis?
8. What is meant by capacity planning?
9. Name different types of production systems.
10. Define TPM.
11. Enlist various factors affecting the location of a heavy manufacturing facility.
12. What is line balancing?
13. What do you mean by JIT?
14. Define work measurement.
15. Identify any four important functions of operations management.
16. Name the important considerations related to product design.
17. Identify any three tangible (quantitative) and any three intangible (qualitative) factors related to location analysis.
18. Define method study.
19. Define sequencing?
20. What is conflict between marketing and production functions.
21. Differentiate between Job rotation and Job enlargement.
22. What are uses of production and operation standards?
23. Define Job design.
24. Define therbligs.
25. Differentiate between PERT and CPM.

IMPORTANT QUESTIONS
(*10 Marks*)

1. Explain the factors, which largely influence the make or buy decision.
2. Describe the scope of production and operation management.
3. Explain the role of production management in a business organisation.
4. (*a*) Differentiate between value and cost.

   (*b*) Explain the terms use value, esteem value, time value and place value.

   (*c*) What are objective of value analysis.
5. In a time study various allowances are given? What are these and how are they arrived at? Do the allowances vary from industry to industry.
6. Suppose you have to establish a milk plant. What factors you will consider to locate it? Discuss.
7. How the Capacity utilization is influenced by seasonal demand and how it improved.
8. What are the main differences between product design and product development, explain using some suitable examples.
9. Explain various techniques of Job design.
10. Explain the relationship between production design and process design.
11. What is operation strategy? Explain the process.
12. Write detailed notes on:

    (*a*) Classifictions and codification of materials.

    (*b*) Product life Cycle.
13. What do you mean by value analysis? Explain the procedure for conducting value analysis. Is there any relationship between productivity improvement and value analysis?
14. Explain Kanban system of JIT.
15. Explain '5S' techniques.
16. What characteristics you think a good plant layout should have? Also describe the various types of layout enlisting the advantages and limitations of each.
17. Describe the different types of production system. What are the advantages and disadvantages of each?
18. Discuss the objective of production scheduling. Explain briefly the various techniques employed for this purpose.
19. Explain six sigma implementation process.
20. Define TPM. What is need of TPM? Differentiate TQM and TPM.

❑❑❑

Index

L

M

N

O

P

R

S

T

V

W

Win Prizes

ATTENTION READERS

Win Prizes

We request for your frank assessment regarding various aspects of the book as given below :

Production and Operations Management

BY : Pratibha Garg **(UPTU)**

Please Fill up the Sheet in Neat and Clean Writing

(i) What topics of your syllabus, important for your examination point of view, are not covered in this book ?

....

....

....

....

(ii) What are the chapters wherein the treatment of the subject matter is not systematic or organised or updated ?

....

....

....

....

(iii) Have you come across misprints/mistakes/factual inaccuracies in the book ? Please specify the chapters and the page numbers.

....

....

....

....

(iv) What other books on the same subject have you found/heard better than the present book ? Please specify in terms of price and quality.

....

....

....

....

(v) Further suggestions and comments for the improvement of this book.

....

....

....

....

(PS : If need be please attach a separate sheet to write your views)

OTHER DETAILS

(i) Who recommended you this book ? Please tick (✓) in the box near the option relevant to you.

☐ Teacher ☐ Friends ☐ Bookseller

....

....

....

....

(ii) Name of the recommending teacher, his designation and address:

....

....

....

....

(iii) Name and address of the bookseller you purchased the book from :

....

....

....

....

(iv) Name and address of your institution (Please mention the University or Board as the case may be)

....

....

....

....

(v) Your name and complete address :

....

....

....

....

(vi) Course you are enrolled in :

....

....

....

....

The best assessment will be awarded each month : The award will be in the form of our publications as decided by the Editorial Board.

Please mail the filled up coupon at your earliest to :

S.K. KATARIA & SONS®
4760-61/23, Ansari Road,
Daryaganj, New Delhi-110002
☎ 23243489, 23269324
Fax : 91-11-23243489
email : katariabooks@yahoo.com